Laboratory Manual

Organic and Biological

Chemistry

Structures of Life

Karen C. Timberlake

Professor Emeritus, Los Angeles Valley College

Benjamin
Cummings

San Francisco • Boston • New York
Capetown • Hong Kong • London • Madrid • Mexico City
Montreal • Munich • Paris • Singapore • Sydney • Tokyo • Toronto

Acquisitions Editor: Maureen Kennedy
Project Editor: Claudia Herman
Managing Editor: Joan Marsh
Marketing Manager: Christy Lawrence
Manufacturing Coordinator: Vivian McDougal
Cover Design: Tony Asaro
Cover Photographs: John Bagley, Richard Tauber
Cover Illustration: Blakeley Kim

Some of the experiments in this lab book may be hazardous if materials are not handled properly or procedures are not followed correctly. Safety procedures of your college must be followed as directed by your instructor. Safety precautions must be utilized when you work with laboratory equipment, glassware, and chemicals.

ISBN 0-8053-2992-7

Benjamin
Cummings

2 3 4 5 6 7 8 9 10–PBT–05 04 03 02

www.aw.com/bc

Preface

Welcome to the *Laboratory Manual for Organic and Biological Chemistry: Structures of Life.* In the process of writing lab manuals, I have developed experiments that illustrate each of the chemical principles we discuss from the first day of class. I have also taken care to make each experiment workable as well as providing critical thinking for the student. The goals of this laboratory manual address the following areas:

1. **Experiments relate to basic concepts of chemistry and health.** Experiments are designed to illustrate the chemical principles we discuss in our classes. They include experiments that relate to health and medicine, and often use common materials that are familiar to students.

2. **Experiments are flexible.** Each experiment includes a flexible group of sections, which allows instructors to select the sections to fit into their weekly laboratory schedule. Lab times and comments are given for each.

3. **Safety.** A detailed safety section in the preface includes a safety quiz. The aim here is to highlight the safety and equipment preparation on the first day of lab. In addition, each lab contains reminders of safety behavior. Students are reminded to wear goggles for every lab session. Some experiments are recommended as instructor demonstrations.

4. **Experiment format provides clear instructions and evaluation.** Each lab begins with a set of goals, a discussion of the topics, and examples of calculations. The report pages begin with pre-lab questions to prepare students for lab work. Students obtain data, draw graphs, make calculations, and write conclusions about their results. Each lab contains questions and problems that require the student to discuss the experiment, make additional calculations, and use critical thinking to apply concepts to real life.

5. **Stockroom preparation of chemicals.** Materials for each experiment are listed in the appendix with amounts given for 20 students working in pairs. Most lab sessions use standard lab equipment and chemicals that are readily available and inexpensive. In some cases students bring samples from home.

I hope that this laboratory manual will help you in your chemistry instruction and that students will find they learn chemistry by participating in the laboratory experience.

Karen C. Timberlake
Los Angeles Valley College
Valley Glen, CA 91401

Here you are in a chemistry laboratory with your laboratory book in front of you. Perhaps you have already been assigned a laboratory drawer, full of glassware and equipment you may never have seen before. Looking around the laboratory, you may see bottles of chemical compounds, balances, burners, and other equipment that you are going to use. This may very well be your first experience with experimental procedures. At this point you may have some questions about what is expected of you. This laboratory manual is written with those considerations in mind.

The activities in this manual were written specifically to parallel the chemistry you are learning in the lecture portion of class. Many of the laboratory activities include materials that may be familiar to you, such as household products, diet drinks, cabbage juice, antacids, and aspirin. In this way, chemical topics are related to the real world and to your own science experience. Some of the labs teach basic skills; others encourage you to extend your scientific curiosity beyond the lab.

It is important to realize that the value of the laboratory experience depends on the time and effort you invest in it. Only then will you find that the laboratory can be a valuable learning experience and an integral part of the chemistry class. The laboratory gives you an opportunity to go beyond the lectures and words in your textbook and experience the scientific process from which conclusions and theories concerning chemical behavior are drawn. In some experiments, the concepts are correlated with health and biological concepts. Chemistry is not an inanimate science, but one that helps us to understand the behavior of living systems.

Using This Laboratory Manual

Each experiment begins with learning goals to give you an overview of the topics you will be studying in that experiment. Each experiment is correlated to concepts you are currently learning in your chemistry class. Your instructor will indicate which activities you are to do. At the beginning of each experiment, you will also find a list of the materials needed for each activity.

The experimental procedures are written to guide you through each laboratory activity. When you are ready to begin the lab, remove the report sheet at the end of the laboratory instructions. Place it next to the procedures for that section. Read and measure carefully, report your data, and follow instructions to complete the necessary calculations. You may also be asked to answer some or all of the follow-up questions and problems designed to test your understanding of the concepts from the experiment.

It is my hope that the laboratory experience will help illuminate the concepts you are learning in the classroom. The experimental process can help make chemistry a real and exciting part of your life and provide you with skills necessary for your future.

Contents

Contents

The chemistry laboratory with its equipment, glassware, and chemicals has the potential for accidents. In order to avoid accidents, precautions must be taken by every student to ensure the safety of everyone working in the laboratory. By following the rules for handling chemicals safely and carrying out only the approved procedures, you will create a safe environment in the laboratory. After you have read the following sections, complete the safety quiz and the questions on laboratory equipment. Then sign and submit the commitment to lab safety.

A. Preparing for Laboratory Work

Pre-read Before you come to the laboratory, read the discussion of and directions for the experiment you will be doing. Make sure you know what the experiment is about before you start the actual work. If you have a question, ask your instructor to clarify the procedures.

Do assigned work only Do only the experiments that have been assigned by your instructor. No unauthorized experiments are to be carried out in the laboratory. Experiments are done at assigned times, unless you have an open lab situation. Your instructor must approve any change in procedure.

Do not work alone in a laboratory.

Safety awareness Learn the location and use of the emergency eyewash fountains, the emergency shower, fire blanket, fire extinguishers, and exits. Memorize their locations in the laboratory. Be aware of other students in the lab carrying chemicals to their desk or to a balance.

 ### *APPROVED EYE PROTECTION IS REQUIRED AT ALL TIMES!*

Safety goggles must be worn all the time you are in the lab The particular type depends on state law, which usually requires industrial-quality eye protection. Contact lenses may be worn in the lab if needed for therapeutic reasons, provided that **safety goggles** are worn over the contact lenses. Contact lenses without goggles are dangerous because splashed chemicals make them difficult to remove. If chemicals accumulate under a lens, permanent eye damage can result. If a chemical should splash into your eyes, flood the eyes with water at the eyewash fountain. Continue to rinse with water for at least 10 minutes.

Wear protective clothing Wear sensible clothing in the laboratory. Loose sleeves, shorts, or open-toed shoes can be dangerous. A lab coat is useful in protecting clothes and covering arms. Wear shoes that cover your feet to prevent glass cuts; wear long pants and long-sleeved shirts to protect skin. Long hair should be tied back so it does not fall into chemicals or a flame from a Bunsen burner.

No food or drink is allowed at any time in the laboratory Do not let your friends or children visit while you are working in the lab; have them wait outside.

Prepare your work area Before you begin a lab, clear the lab bench or work area of all your personal items, such as backpacks, books, sweaters, and coats. Find a storage place in the lab for them. All you will need is your laboratory manual, calculator, pen or pencil, text, and equipment from your lab drawer.

B. Handling Chemicals Safely

Check labels twice Be sure you take the correct chemical. *DOUBLE-CHECK THE LABEL* on the bottle before you remove a chemical from its container. For example, sodium sulfate (Na_2SO_4) could be mistaken for sodium sulfite (Na_2SO_3) if the label is not read carefully.

Use small amounts of chemicals Pour or transfer a chemical into a small, clean container (beaker, test tube, flask, etc.) available in your lab drawer. To avoid contamination of the chemical reagents, never insert droppers, pipets, or spatulas into the reagent bottles. Take only the quantity of chemical you need for the experiment. Do not keep a reagent bottle at your desk; *return* it to its proper location in the laboratory. Label the container. Many containers have etched sections on which you can write in pencil. If not, use tape or a marking pencil.

Do not return chemicals to the original containers To avoid contamination of chemicals, dispose of used chemicals according to your instructor's instructions. *Never return unused chemicals to reagent bottles.* Some liquids and water-soluble compounds may be washed down the sink with plenty of water, but check with your instructor first. Dispose of organic compounds in specially marked containers in the hoods.

Do not taste chemicals; smell a chemical cautiously Never use any equipment in the drawer such as a beaker to drink from. When required to note the odor of a chemical, first take a deep breath of fresh air and hold it while you use your hand to fan some vapors toward your nose and note the odor. Do not inhale the fumes directly. If a compound gives off an irritating vapor, use it in the fume hood to avoid exposure.

Do not shake laboratory thermometers Laboratory thermometers respond quickly to the temperature of their environment. Shaking a thermometer is unnecessary and can cause breakage.

Liquid spills Spills of water or liquids at your work area or floor should be cleaned up immediately. Small spills of liquid chemicals can be cleaned up with a paper towel. Large chemical spills must be treated with absorbing material such as cat litter. Place the contaminated material in a waste disposal bag and label it. If a liquid chemical is spilled on the skin, flood *immediately with water* for at least 10 minutes. Any clothing soaked with a chemical must be removed immediately because an absorbed chemical can continue to damage the skin.

Mercury spills The cleanup of mercury requires special attention. Mercury spills may occur from broken thermometers. Notify your instructor immediately of any mercury spills so that special methods can be used to clean up the mercury. Place any free mercury and mercury cleanup material in special containers for mercury only.

Laboratory accidents Always notify your instructor of any chemical spill or accident in the laboratory. Broken glass can be swept up with a brush and pan and placed in a specially labeled container for broken glass. Cuts are the most common injuries in a lab. If a cut should occur, wash, elevate, and apply pressure if necessary. Always inform your instructor of any laboratory accident.

Clean up Wash glassware as you work. Begin your cleanup 15 minutes before the end of the laboratory session. Return any borrowed equipment to the stockroom. Be sure that you always turn off the gas and water at your work area. Make sure you leave a clean desk. Check the balance you used. *Wash your hands before you leave the laboratory.*

C. Heating Chemicals Safely

Heat only heat-resistant glassware Only glassware marked Pyrex® or Kimax® can be heated; other glassware may shatter. To heat a substance in a test tube, use a test tube holder. Holding the test tube at an angle, move it continuously through the flame. Never point the open end of the test tube at anyone or look directly into it. A hot piece of iron or glass looks the same as it does at room temperature. Place a hot object on a tile or a wire screen to cool.

Flammable liquids Never heat a flammable liquid over an open flame. If heating is necessary, your instructor will indicate the use of a steam bath or a hot plate.

Never heat a closed container When a closed system is heated, it can explode as pressure inside builds.

Fire Small fires can be extinguished by covering them with a watch glass. If a larger fire is involved, use a fire extinguisher to douse the flames. *Do not direct a fire extinguisher at other people in the laboratory.* Shut off gas burners in the laboratory. When working in a lab, tie long hair back away from the face. If someone else's clothing or hair catches on fire, get them to the floor and roll them into a fire blanket. They may also be placed under the safety shower to extinguish flames. Cold water or ice may be applied to small burns.

D. Waste Disposal

As you work in the laboratory, chemical wastes are produced. Although we will use small quantities of materials, some waste products are unavoidable. To dispose of these chemical wastes safely, you need to know some general rules for chemical waste disposal.

Metals Metals should be placed in a container to be recycled.

Nonhazardous chemical wastes Substances such as sodium chloride (NaCl) that are soluble in water and are not hazardous may be emptied into the sink. If the waste is a solid, dissolve it in water before disposal.

Hazardous chemical wastes If a substance is hazardous or not soluble in water, it must be placed in a container that is labeled for waste disposal. Your instructor will inform you if chemical wastes are hazardous and identify the proper waste containers. *If you are not sure about the proper disposal of a substance, ask your instructor.* The labels on a waste container should indicate if the contents are hazardous, the name of the chemical waste, and the date that the container was placed in the lab.

Hazard rating The general hazards of a chemical are presented in a spatial arrangement of numbers with the flammability rating at twelve o'clock, the reactivity rating at the three o'clock position, and the health rating at the nine o'clock position. At the six o'clock position, information may be given on the reactivity of the substance with water. If there is unusual reactivity with water, the symbol 𝖶 (do not mix with water) is shown. In the laboratory, you may see these ratings in color with blue for health hazard, red for flammability, and yellow for reactivity hazards.

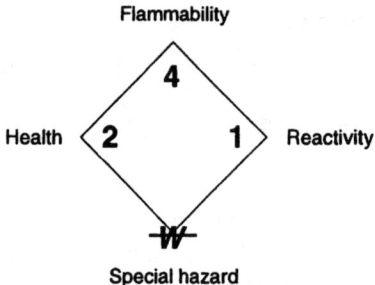

A chemical is assigned a relative hazard rating that ranges from 1 (little hazard) to 4 (extreme hazard). The health hazard indicates the likelihood that a material will cause injury due to exposure by contact, inhalation, or ingestion. The flammability hazard indicates the potential for burning. The reactivity hazard indicates the instability of the material by itself or with water with subsequent release of energy. Special hazards may be included such as 𝖶 for reactivity with water or OX for oxidizing properties.

E. Safety Quiz

The safety quiz will review the preceding safety discussion. Circle the correct answer(s) in each of the following questions. Check your answers on page xiii.

1. Approved eye protection is to be worn
    a. For certain experiments
    b. Only for hazardous experiments
    c. All the time

2. Eating in the laboratory is
    a. Not permitted
    b. Allowed at lunch time
    c. All right if you are careful

3. If you need to smell a chemical, you should
    a. Inhale deeply over the test tube
    b. Take a breath of air and fan the vapors toward you
    c. Put some of the chemical in your hand, and smell it

4. When heating liquids in a test tube, you should
    a. Move the tube back and forth through the flame
    b. Look directly into the open end of the test tube to see what is happening
    c. Direct the open end of the tube away from other students

5. Unauthorized experiments are
    a. All right as long as they don't seem hazardous
    b. All right as long as no one finds out
    c. Not allowed

6. If a chemical is spilled on your skin, you should
    a. Wait to see if it stings
    b. Flood the area with water for 10 minutes
    c. Add another chemical to absorb it

7. When taking liquids from a reagent bottle,
    a. Insert a dropper
    b. Pour the reagent into a small container
    c. Put back what you don't use

8. In the laboratory, open-toed shoes and shorts are
    a. Okay if the weather is hot
    b. All right if you wear a lab apron
    c. Dangerous and should not be worn

9. When is it all right to taste a chemical?
    a. Never
    b. When the chemical is not hazardous
    c. When you use a clean beaker

10. After you use a reagent bottle,
    a. Keep it at your desk in case you need more
    b. Return it to its proper location
    c. Play a joke on your friends and hide it

11. Before starting an experiment,
    a. Read the entire procedure
    b. Ask your lab partner how to do the experiment
    c. Skip to the laboratory report and try to figure out what to do

12. Working alone in the laboratory without supervision is
    a. All right if the experiment is not too hazardous
    b. Not allowed
    c. Allowed if you are sure you can complete the experiment without help

13. You should wash your hands
    a. Only if they are dirty
    b. Before eating lunch in the lab.
    c. Before you leave the lab

14. Personal items (books, sweater, etc.) should be
    a. Kept on your lab bench
    b. Left outside
    c. Stored out of the way, not on the lab bench

15. When you have taken too much of a chemical, you should
    a. Return the excess to the reagent bottle
    b. Store it in your lab locker for future use
    c. Discard it using proper disposal procedures

16. In the lab, you should wear
    a. Practical, protective clothing
    b. Something fashionable
    c. Shorts and loose-sleeved shirts

17. If a chemical is spilled on the table,
    a. Clean it up right away
    b. Let the stockroom help clean it up
    c. Use appropriate adsorbent if necessary

18. If mercury is spilled,
    a. Pick it up with a dropper
    b. Call your instructor
    c. Push it under the table where no one can see it

19. If your hair or shirt catches on fire, you should
    a. Use the safety shower to extinguish the flames
    b. Drop to the ground and roll
    c. Use the fire blanket to put it out

20. A hazardous waste should be
    a. Placed in a special waste container
    b. Washed down the drain
    c. Placed in the wastebasket

Answer Key to Safety Quiz

1. c	2. a	3. b	4. a, c	5. c	6. b	7. b
8. c	9. a	10. b	11. a	12. b	13. c	14. c
15. c	16. a	17. a, c	18. b	19. a, b, c	20. c	

Graphing Experimental Data

When a group of experimental quantities is determined, a graph can be prepared that gives a pictorial representation of the data. After a data table is prepared, a series of steps is followed to construct a graph.

Preparing a Data Table

A data table is prepared from measurements. Suppose we measured the distance traveled in a given time by a bicycle rider. Table 1 is a data table prepared by listing the two variables, time and distance, that we measured.

Table 1 *Time and Distance Measurement*

Time (hr)	Distance (km)
1	5
3	14
4	20
6	30
7	33
8	40
9	46
10	50

Constructing the Graph

Draw vertical and horizontal axes Draw a vertical and a horizontal axis on the appropriate graph paper. The lines should be set in to leave a margin for numbers and labels, but the graph should cover most of the graph paper. Place a title at the top of the graph. The title should describe the quantities that will be placed on the axes.

Label each axis The label for each axis reflects the measurement listed in the data table. On our sample graph, the labels are time (hr) for the horizontal axis and distance (km) for the vertical axis.

Apply constant scales On each axis apply a scale of equal intervals that includes the full range of data points (low to high) you have in the data table. The intervals on a scale must be *equally spaced* and fit on the line you have drawn. Do not exceed the graph lines. Use intervals on each axis that are convenient counting units (2, 4, 6, 8, etc. or 5, 10, 15, etc.). The interval size on one axis does not need to match the size of the intervals on the other axis.

For our sample graph, we used a scale for a distance range of 0 km to 50 km. Each graph division represents 5 km. (You only have to number a few lines in order to interpret the scale. It gets too crowded with numbers if every line is marked.) Every two divisions on the time scale represent a time interval of 1 hour within the 10-hour time span of the bicycle ride. See Figure 1.

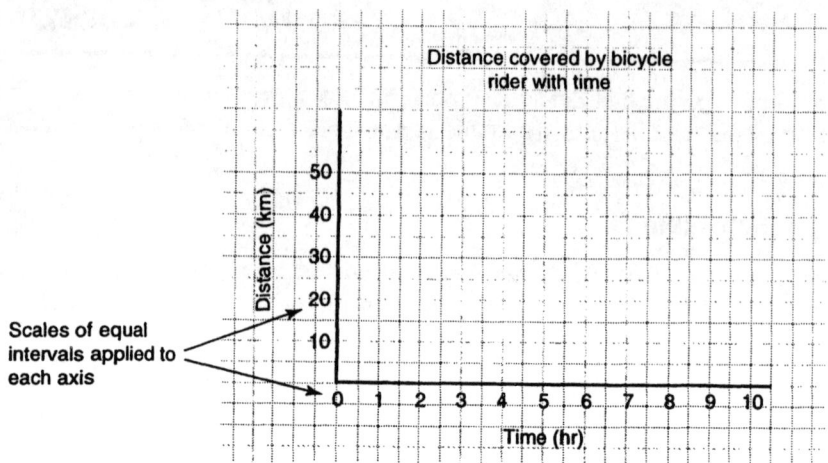

Figure 1   Marking equal intervals for distance and time on the axes

Plot the data points Plot the points for each pair of measurements on the data table. Follow a measurement on the vertical axis across until it meets a line that would be drawn from the corresponding measured value on the horizontal axis.

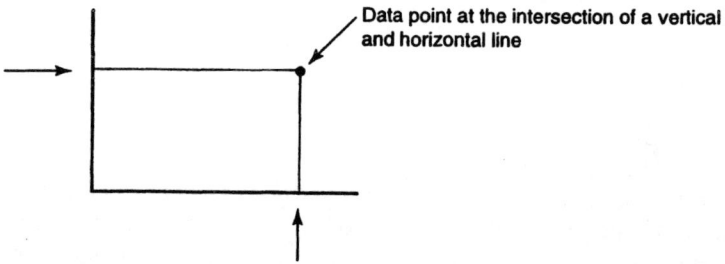

For example, at 4 hours, the rider has traveled 20 km. On the graph, find 20 km on the distance scale, and 4 hr on the time scale. Then follow the perpendicular lines to where they intersect. That is a point on the graph. Plotting each data pair will show the relationship between distance and time. A smooth line or curve is drawn that best fits the data points. However, some points may not fit on the line or curve you draw. That occurs when error is associated with the measurements or when the data are affected by other variables, such as terrain and energy level for the bicycle rider. See Figure 2.

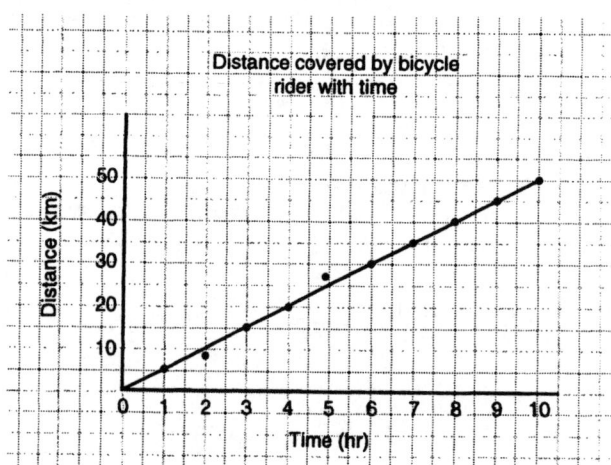

Figure 2   A completed graph with data points connected in a smooth line

Using the Laboratory Burner

Wear your goggles!

Materials: Bunsen burner, striker or matches

In the laboratory, substances are often heated with a Bunsen burner, shown in the figure below. The burner consists of a metal tube and base connected to a gas source. The flow of gas is controlled by adjusting the gas lever at the bench or by turning the wheel at the base of the burner. The amount of air that enters the burner is adjusted by twisting the tube to open or close the air vents. The gas and air mixture is ignited at the top of the tube using a match or a striker. Make sure the gas valves are tightly closed when you leave the laboratory after using the Bunsen burner.

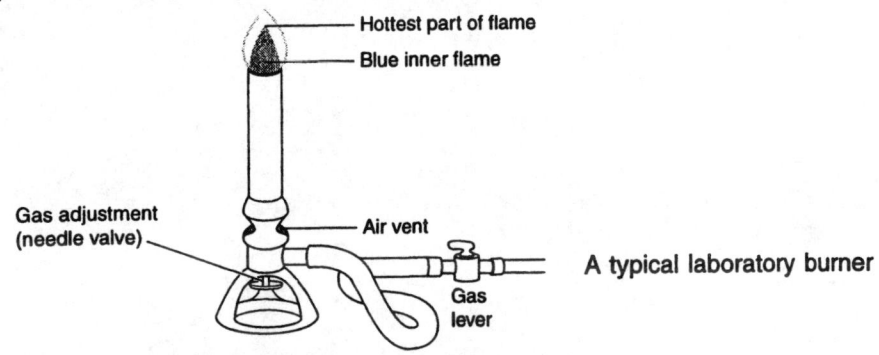

A typical laboratory burner

1. Before you light the burner, practice the following:
   a. Open and close the gas lever at the lab bench.
   b. Open and close the gas needle valve (wheel) at the base of the burner.
   c. Open and close the air vents.

   With the air vents closed, ignite the burner with a striker or match. Your instructor may demonstrate the use of the striker. Turn on the gas and hold the flame or spark at the top rim of the burner tube.

2. If the flame is yellow and sooty, the gas mixture does not have an adequate supply of oxygen. Open the air vents until the color of the flame changes to blue. Adjust the gas flow until you have a flame that is 6–8 cm high with two distinct parts, an inner cone and an outer flame.

3. The hottest part of a flame is at the tip of the inner blue flame. For the most effective heating, be sure that the tip of the inner flame is placed just under the substance you heat. *Remember what you heated: Hot metal and glass items do not look hot!*

Questions

Q.1 What is the color of the flame with the air vent closed? _____

Open? _____

How do you control the height of the flame?_____

_____

Q.2 What is the appearance of a flame that is used for heating?_____

_____

Q.3 Where is the hottest part of a flame?_____

Properties of Organic Compounds

Goals

- Observe chemical and physical properties of organic and inorganic compounds.
- Identity functional groups in three-dimensional models.

Discussion

A. Color, Odor, and Physical State

Organic compounds are made of carbon and hydrogen, and sometimes oxygen and nitrogen. Of all the elements, only carbon atoms bond to many more carbon atoms, a unique ability that gives rise to many more organic compounds than all the inorganic compounds known today. The covalent bonds in organic compounds and the ionic bonds in inorganic compounds account for several of the differences we will observe in their physical and chemical properties. See Table 21.1.

Table 21.1 *Comparing Some Properties of Organic and Inorganic Compounds*

Organic Compounds	Inorganic Compounds
Covalent bonds	Ionic or polar bonds
Soluble in nonpolar solvents, not water	Soluble in water
Low melting and boiling points	High melting and boiling points
Strong, distinct odors	Usually no odor
Poor or nonconductors of electricity	Good conductors of electricity
Flammable	Not flammable

B. Solubility

Typically, inorganic compounds that are ionic are soluble in water, a polar compound, but organic compounds are nonpolar and thus are not soluble in water. However, organic compounds are soluble in organic solvents because they are both nonpolar. A general rule for solubility is that "like dissolves like."

C. Combustion

Many organic compounds react with oxygen, a reaction called *combustion,* to form carbon dioxide and water. Combustion is the reaction that occurs when gasoline burns with oxygen in the engine of a car or when natural gas, methane, burns in a heater or stove. In a combustion reaction, heat is given off; the reaction is exothermic. Equations for the combustion of methane and propane are written as follows:

$$CH_4(g) + 2O_2(g) \longrightarrow CO_2(g) + 2H_2O(g) + heat$$
methane

$$C_3H_8(g) + 5O_2(g) \longrightarrow 3CO_2(g) + 4H_2O(g) + heat$$
propane

D. Functional Groups

Although there are millions of organic compounds, they can be classified according to organic families. Each family contains a characteristic structural feature called a functional group, which is a certain atom or group of atoms that give similar physical and chemical properties to that family. Because the organic compounds in a family contain the same functional group, they undergo the same types of chemical reactions. In this lab, we will take a look at some of the common functional groups that allow us to classify organic compounds according to the structure,

Alkanes, alkenes, and alkynes are hydrocarbons that consist of only carbon and hydrogen atoms. Alkanes contain carbon-carbon single bonds, whereas alkenes contain one or more carbon-carbon double bonds, and alkynes contain a carbon-carbon triple bond. To write a condensed structural formula, the hydrogen atoms attached to each carbon are written adjacent to the symbol C for carbon. Thus, a CH_3- is the abbreviation for a carbon attached to three hydrogen atoms, whereas $-CH_2-$ shows a carbon attached to two hydrogen atoms.

H H	H H	
| |	| |	
H—C—C—H	H—C=C—H	H—C≡C—H
| |		
H H		
An alkane	An alkene	An alkyne
CH_3—CH_3	CH_2=CH_2	HC≡CH

Condensed
Structural
Formulas

Alcohols and ethers contain an oxygen atom. Alcohols have a *hydroxyl group*, which is an –OH group, bonded to a carbon atom. In an ether, the oxygen atom is bonded to two carbon atoms.

H H	H H
| |	| |
H—C—C—O—H	H—C—O—C—H
| |	| |
H H	H H
An alcohol	An ether
CH_3—CH_2—OH	CH_3—O—CH_3

Condensed
Structural
Formulas

Aldehydes and ketones, contain a carbonyl group, which is a carbon-oxygen double bond (C=O). In an aldehyde, the carbon bonds to at least one hydrogen atom. In a ketone, the carbon in the carbonyl group bonds to two other carbon atoms.

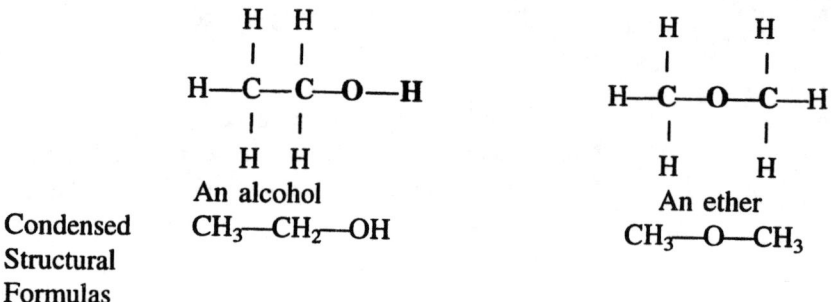

O	O	O
‖	‖	‖
H—C—H	CH_3—C—H	H_3C—C—CH_3
An aldehyde	An aldehyde	A ketone

Carboxylic acids and esters contain the functional group is the carboxyl group, which is a combination of a carbonyl and hydroxyl group. In a carboxylic acid, the oxygen is bonded to a hydrogen atom, whereas in an ester the oxygen is bonded to a carbon and not to hydrogen.

$$
\begin{array}{c}
\overset{\displaystyle O}{\overset{\displaystyle \|}{CH_3-C-O-H}}
\end{array}
\qquad \text{or} \qquad CH_3COOH \qquad \text{or} \qquad CH_3CO_2H
$$

A carboxylic acid

$$
\begin{array}{c}
\overset{\displaystyle O}{\overset{\displaystyle \|}{CH_3-C-O-CH_3}}
\end{array}
\qquad \text{or} \qquad CH_3COOCH_3 \qquad \text{or} \qquad CH_3CO_2CH_3
$$

An ester

Amines contain a nitrogen atom because they are derivatives of ammonia, NH_3. In an amine, one or more carbon groups replace the hydrogen atoms in ammonia. Amines are classified as primary, secondary, or tertiary according to the number of carbon groups bonded to the nitrogen atoms.

$$
NH_3 \qquad CH_3-NH_2 \qquad
\begin{array}{c} CH_3-NH \\ | \\ CH_3 \end{array}
\qquad
\begin{array}{c} CH_3-N-CH_3 \\ | \\ CH_3 \end{array}
$$

Ammonia   A primary   A secondary   A tertiary
amine (1°)   amine (2°)   amine (3°)

Lab Information

Time: 2 hr

Comments: Tear out the report sheets and place them next to the matching procedures. ***Organic compounds are extremely flammable! Use of the Bunsen burner is prohibited.***

Related Topics: Organic compounds, hydrocarbons, solubility, combustion, complete structural formula, functional groups

Experimental Procedures   *Wear your safety goggles!*

A.   Color, Odor, and Physical State *(This may be a lab display.)*

Materials: Test tubes (6), test tube rack, spatulas, NaCl(*s*), KI(*s*), toluene, benzoic acid, cyclohexane, water, chemistry handbook

Place each substance into a separate test tube: a few crystals of NaCl, KI, and benzoic acid, and 10 drops each of cyclohexane, toluene, and water. Or if a display is available, observe the samples in a test tube rack in the hood. Record the formula, physical state (solid, liquid, or gas), and odor of each one. To check for odor, first take a breath and hold it while you gently fan the air above the test tube toward you. Look up the melting point of each compound using a chemistry handbook. Record. State the types of bonds in each as ionic or covalent. Identify each as an organic or inorganic compound.

B. **Solubility** *(This may be a demonstration or lab display.)*

Materials: Test tubes, spatulas, NaCl(*s*), toluene, cyclohexane

Work in the hood: Be sure to work with the compounds such as cyclohexane in ventilation hoods, and then dispose of them in the proper waste containers. Place 10 drops of cyclohexane and 10 drops of water in a test tube. Record your observations. Identify the upper layer and the lower layer.

Place a few crystals of NaCl in one test tube and 10 drops of toluene in another test tube. To each sample, add 15 drops of cyclohexane, a nonpolar solvent. Shake gently or tap the bottom of the test tube to mix. Record whether each substance is soluble (S) or insoluble (I) in cyclohexane.

Repeat the experiment with the two substances, but this time add 15 drops of water, a polar solvent. Record whether each substance is soluble (S) or insoluble (I) in water. Identify each substance as an organic or inorganic compound.

> **Dispose of organic substances in the proper waste container.**

C. **Combustion** *(This may be a demonstration by your instructor.)*

Materials: 2 evaporating dishes, spatulas, wood splints, NaCl(*s*), cyclohexane

Work in the hood: Place a small amount (pea-size) of NaCl in an evaporating dish set in an iron ring. Ignite a splint and hold the flame to the NaCl. Record whether the substance burns. Repeat the experiment using 5 drops of cyclohexane instead of NaCl. If the substance burns, note the color of the flame. Identify each as an organic or inorganic compound.

D. **Functional Groups**

Materials: An organic model kit or prepared models of organic compounds

D.1 Observe models of organic compounds listed in the table. Or using an organic model kit, construct ball-and-stick models of each of the compounds. Place wooden dowels or springs in all the holes in the carbon atom (black) and attach hydrogen (yellow) atoms, oxygen atoms (red), or nitrogen atoms (blue) as required to complete the functional group for each. Draw a full structural formula showing all the bonds to each carbon atom.

D.2 Circle the functional group in each structure. Classify the organic compound according to the functional group

Report Sheet - Lab 21

Date _____   Name _____

Section _____   Team _____ _____

Instructor _____

Pre-Lab Study Questions

1. Would you expect an organic compound to be soluble in water? Why?

2. Which is more flammable: an organic or inorganic compound?

A. Color, Odor, and Physical State

Name	Formula	Physical State	Odor	Melting Point	Type of Bonds?	Organic or Inorganic?
Sodium chloride						
Cyclohexane	C_6H_{12}					
Potassium iodide						
Benzoic acid	$C_7H_6O_2$					
Toluene	C_7H_8					
Water						

B. Solubility

In the mixture, water is the _____ layer and cyclohexane is the _____ layer.

Solute	Solubility in Cyclohexane	Solubility in Water	Organic or Inorganic?
NaCl			
Toluene			

Report Sheet - Lab 21

C. Combustion

Compound	Flammable (Color of Flame)	Not Flammable	Organic or Inorganic?
NaCl			
Cyclohexane			

From your observations of the chemical and physical properties of alkanes as organic compounds, complete the following table:

Property	Organic Compounds	Inorganic Compounds
Elements		
Bonding		
Melting points		
Strong odors		
Flammability		
Solubility		

Questions and Problems

Q.1 Describe three properties you can use to distinguish between organic and inorganic compounds.

Q.2 A white solid has no odor, is soluble in water, and is not flammable. Would you expect it to be an organic or an inorganic substance? Why?

Q.3 A clear liquid with a gasoline-like odor forms a layer when added to water. Would you expect it to be an organic or an inorganic substance? Why?

Report Sheet - Lab 21

D.   Functional Groups

Compound	D.1 Full Structural Formula	D.2 Organic Family
CH_3—OH		
CH_3—CH_2—CH_3		
CH_2=CH_2		
CH_3—O—CH_3		
CH_3—NH_2		
CH_3—$\overset{\overset{\textstyle O}{\|\|}}{C}$—OH		
CH_3—$\overset{\overset{\textstyle O}{\|\|}}{C}$—$CH_3$		
CH_3—$\overset{\overset{\textstyle H}{\|}}{N}$—$CH_3$		

Report Sheet - Lab 21

Questions and Problems

Q.4 Classify the following organic compounds according to their functional groups:

a. _____ $CH_3\!-\!CH_2\!-\!CH\!=\!CH\!-\!CH_3$

b. _____ $CH_3\!-\!CH_2\!-\!\overset{\displaystyle \overset{\textstyle H}{|}}{N}\!-\!CH_2\!-\!CH_3$

c. _____ $CH_3\!-\!CH_2\!-\!O\!-\!CH_3$

d. _____ $CH_3\!-\!\overset{\displaystyle \overset{\textstyle O}{\|}}{C}\!-\!CH_2\!-\!CH_3$

e. _____ $CH_3\!-\!CH_2\!-\!CH_2\!-\!\overset{\displaystyle \overset{\textstyle O}{\|}}{C}\!-\!OH$

f. _____ $CH_3\!-\!CH_2\!-\!CH_2\!-\!\overset{\displaystyle \overset{\textstyle O}{\|}}{C}\!-\!O\!-\!CH_3$

Goals

- Draw formulas for alkanes from their three-dimensional models.
- Write the names of alkanes from their structural formulas.
- Construct models of isomers of alkanes.
- Write the structural formulas for cycloalkanes and haloalkanes.

Discussion

A. Structures of Alkanes

The saturated hydrocarbons represent a group of organic compounds composed of carbon and hydrogen. Alkanes and cycloalkanes are called *saturated* hydrocarbons because their carbon atoms are connected by only single bonds. In each type of alkane, each carbon atom has four valence electrons and must always have four single bonds.

To learn more about the three-dimensional structure of organic compounds, it is helpful to build models using a ball-and-stick model kit. In the kit are wooden (or plastic) balls, which represent the typical elements in organic compounds. Each wooden atom has the correct number of holes drilled for bonds that attach to other atoms. See Table 22.1.

Table 22.1 *Elements and Bonds Represented in the Organic Model Kit*

Color	Element	Number of Bonds
Black	carbon	4
Yellow	hydrogen	1
Red	oxygen	2
Green	chlorine	1
Orange	bromine	1
Purple	iodine	1
Blue	nitrogen	3
Bonds		
Sticks, springs		

The first model to build is methane, CH_4, a hydrocarbon consisting of one carbon atom and four hydrogen atoms. The model of methane shows the three-dimensional shape, a tetrahedron, around a carbon atom.

Complete structural formula   Condensed formula   Geometric formula

To represent this model on paper, its shape is flattened, and the carbon atom is shown attached to four hydrogen atoms. This type of formula is called a *complete structural formula*. However, it is more convenient to use a shortened version called a *condensed structural formula*. To write a condensed formula, the hydrogen atoms are grouped with their carbon atom. The number of hydrogen atoms is written as a subscript. The complete structural formula and the condensed structural formula for C_2H_6 are shown below:

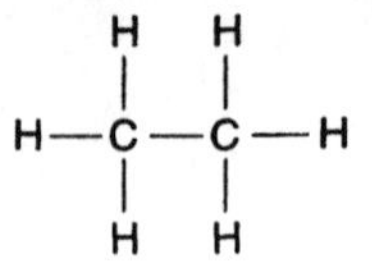

Complete structural formula   Condensed structural formula

Names of Alkanes

The names of alkanes all end with -*ane*. The names of organic compounds are based on the names of the alkane family. See Table 22.2.

Table 22.2   *Names and Formulas of the First Ten Alkanes*

Name	Formula	Name	Formula
Methane	CH_4	Hexane	$CH_3CH_2CH_2CH_2CH_2CH_3$
Ethane	CH_3CH_3	Heptane	$CH_3CH_2CH_2CH_2CH_2CH_2CH_3$
Propane	$CH_3CH_2CH_3$	Octane	$CH_3CH_2CH_2CH_2CH_2CH_2CH_2CH_3$
Butane	$CH_3CH_2CH_2CH_3$	Nonane	$CH_3CH_2CH_2CH_2CH_2CH_2CH_2CH_2CH_3$
Pentane	$CH_3CH_2CH_2CH_2CH_3$	Decane	$CH_3CH_2CH_2CH_2CH_2CH_2CH_2CH_2CH_2CH_3$

B.   Constitutional Isomers

Constitutional Isomers are present when a molecular formula can represent two or more different structural (or condensed) formulas. One structure cannot be converted to the other without breaking and forming new bonds. The isomers have different physical and chemical properties. One of the reasons for the vast array of organic compounds is the phenomenon of isomerism.

Isomers of C_4H_{10}

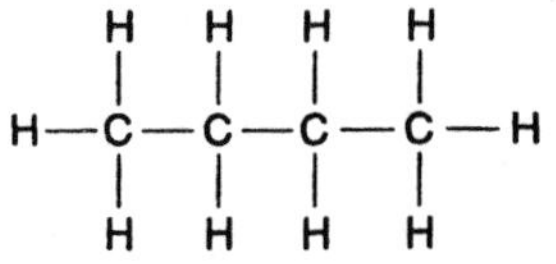

Butane 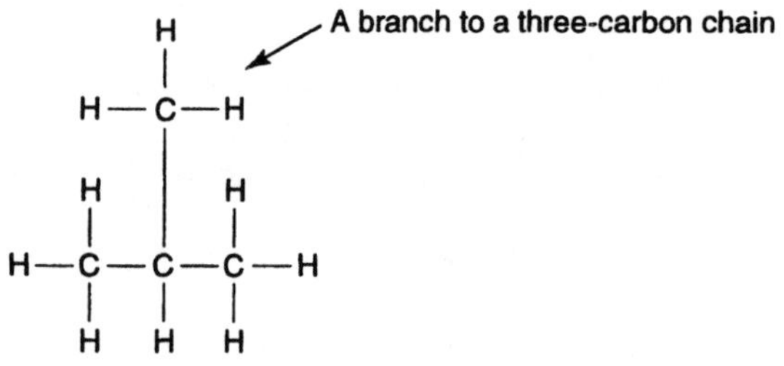   2-Methylpropane

C. Cycloalkanes

In a cycloalkane, an alkane has a cyclic or ring structure. There are no end carbon atoms. The structural formula of a cycloalkane indicates all of the carbon and hydrogen atoms. The condensed formula groups the hydrogen atoms with each of the carbon atoms. Another type of notation called the *geometric* structure is often used to depict a cycloalkane by showing only the bonds that outline the geometric shape of the compound. For example, the geometric shape of cyclopropane is a triangle, and the geometric shape of cyclobutane is a square. Examples of the various structural formulas for cyclobutane are shown below.

Complete structural formula   Condensed formula   Geometric formula

D. Haloalkanes

In a haloalkane, a halogen atom such as chlorine (Cl) or bromine (Br) replaces a hydrogen atom of an alkane or a cycloalkane.

Complete Structural Formula	Condensed Formula	Name
H—C—Cl (with H above and below)	CH_3Cl	Chloromethane (methyl chloride)
Br—C—C—Br (with H atoms)	$BrCH_2CH_2Br$	1,2-Dibromoethane

Lab Information

Time: 2 hr
Comments: Tear out the report sheets and place them next to the matching procedures.
Related Topics: Alkane, cycloalkane, haloalkane, complete structural formula, condensed structural formula, constitutional isomers, naming alkanes

Experimental Procedures

A. Structures of Alkanes

Wear your safety goggles!

Materials: Organic model kit

A.1 Using an organic model kit, construct a ball-and-stick model of a molecule of methane, CH_4. Place wooden dowels in all the holes in the carbon atom (black). Attach hydrogen (yellow) atoms to each. Draw the three-dimensional (tetrahedral) shape of methane. Write the complete structural formula and the condensed structural formula of methane.

213

A.2 Make a model of ethane, C_2H_6. Observe that the tetrahedral shape is maintained for each carbon atom in the structure. Write the complete structural and condensed structural formulas for ethane.

A.3 Make a model of propane, C_3H_8. Write the complete structural and condensed structural formulas for propane.

B. Constitutional Isomers

Materials: Organic model kit, chemistry handbook

B.1 The molecular formula of butane is C_4H_{10}. Construct a model of butane by connecting four carbon atoms in a chain. Draw its complete and condensed structural formulas.

B.2 Make an isomer of C_4H_{10}. Remove an end -CH_3 group and attach it to the center carbon atom. Complete the end of the chain with a hydrogen atom. Write its complete and condensed structural formulas.

B.3 Obtain a chemistry handbook. For each isomer, find the molar mass, melting point, boiling point, and density.

B.4 Make models of the three isomers of C_5H_{12}. Make the continuous-chain isomer first. Draw the complete structural and condensed structural formulas for each. Name each isomer.

B.5 Obtain a chemistry handbook, and find the molar mass, melting point, boiling point, and density of each structural isomer.

C. Cycloalkanes

Materials: Organic model kit

C.1 Use the springs in the model kits to make a model of the cycloalkane with three carbon atoms. Write the complete structural and condensed structural formulas. Draw the geometric formula and name this compound.

C.2 Use the springs in the model kits to make models of a cycloalkane with four carbon atoms, and one with five carbon atoms. Draw the complete structural and condensed structural formulas. Draw the geometric formula and give the name for each.

D. Haloalkanes

Materials: Organic model kit

D.1 Make a model of chloromethane using a green wooden ball for chlorine. Draw the complete structural and condensed structural formulas.

D.2 Make a model of 1,2-dibromoethane using orange wooden balls for the bromine atoms. Draw the complete structural and condensed structural formulas.

D.3 Make a model of 2-iodopropane using a violet wooden ball for iodine. Draw the complete structural and condensed structural formulas.

D.4 Prepare models of four isomers of dichloropropane. Draw the complete structural and condensed structural formulas. Name each isomer.

Report Sheet - Lab 22

Date _____   Name _____

Section _____   Team _____

Instructor _____

Pre-Lab Study Questions

1. What elements are present in alkanes?

2. How does a complete structural formula differ from a condensed structural formula?

3. If isomers of an alkane have the same molecular formula, how do they differ?

A. Structures of Alkanes

A.1 Structure of methane		
Tetrahedral shape	Complete structural formula	Condensed structural formula

A.2 Structure of ethane	
Complete structural formula	Condensed structural formula

A.3 Structure of propane	
Complete structural formula	Condensed structural formula

Report Sheet - Lab 22

Questions and Problems

Q.1 Write the correct name of the following alkanes:

a. $CH_3CH_2CH_3$ _____

$$\begin{array}{c} CH_3 \\ | \end{array}$$

b. $CH_3CH_2CHCHCH_3$ _____

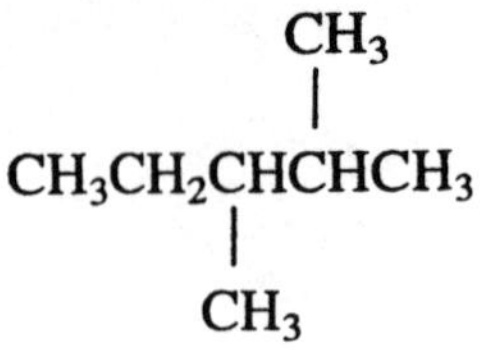

$$\begin{array}{c} | \\ CH_3 \end{array}$$

$$\begin{array}{cc} CH_3 & CH_3 \\ | & | \end{array}$$

c. $CH_3{-}CH{-}CH{-}CH_3$ _____

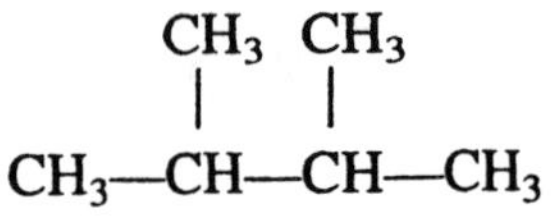

Q.2 Write the condensed formulas for the following:

a. hexane   b. 2,3-dimethylpentane

B.   Constitutional Isomers

B.1 Butane C_4H_{10}	
Complete structural formula	Condensed structural formula
B.2 2-Methylpropane	
Complete structural formula	Condensed structural formul

Report Sheet - Lab 22

B.3 Physical Properties of Isomers of C_4H_{10}				
Isomer	**Molar Mass**	**Melting Point**	**Boiling Point**	**Density**
Butane				
2-Methylpropane (isobutane)				

Questions and Problems

Q.3   In B.3, what physical property is identical for the two isomers of C_4H_{10}?

Q.4   What physical properties are different for the isomers? Explain.

Report Sheet - Lab 22

B.4 Isomers of C_5H_{12}	
Complete structural formula	Condensed structural formula
Name:	
Complete structural formula	Condensed structural formula
Name:	
Complete structural formula	Condensed structural formula
Name:	

B.5 Physical Properties of Isomers of C_5H_{12}				
Isomer	**Molar Mass**	**Melting Point**	**Boiling Point**	**Density**
Pentane				
2-Methylbutane				
2,2-Dimethylpropane				

Report Sheet - Lab 22

Questions and Problems

Q.5 Write the condensed formulas for the five isomers of C_6H_{14}.

C. Cycloalkanes

Complete Structural Formula	Condensed Structural Formula	Geometric Formula
C.1 Three carbon atoms		
Name:		
C.2 Four carbon atoms		
Name:		
Five carbon atoms		
Name:		

Report Sheet - Lab 22

D. Haloalkanes

Complete Structural Formula	Condensed Structural Formula
D.1 Chloromethane	
D.2 1,2-Dibromoethane	
D.3 2-Iodopropane	

Report Sheet - Lab 22

D.4 Four isomers of dichloropropane

Complete Structural Formula	Condensed Structural Formula
Name:	
Name:	
Name:	
Name:	

Report Sheet - Lab 22

Questions and Problems

Q.6 Write the condensed structural formulas and names for all the constitutional isomers with the formula C_4H_9Cl.

Reactions of Hydrocarbons

Goals

- Observe the reactions of hydrocarbons with oxygen, bromine, and potassium permanganate.
- Use chemical tests to distinguish alkanes from alkenes.
- Draw the products of combustion, addition, and/or substitution reactions of alkanes and alkenes.

Discussion

A. Types of Hydrocarbons

Alkanes are saturated hydrocarbons, containing single bonds between carbon atoms. The alkenes and alkynes are unsaturated hydrocarbons, containing double or triple bonds. The double or triple bond, which is unsaturated, is a very reactive site in an alkene or alkyne. The aromatic compounds are hydrocarbons with a benzene ring. See Table 23.1.

Table 23.1 *Bonding Characteristics of Hydrocarbon Families*

Alkane	Alkene	Alkyne	Aromatic
Single bond	Double bond	Triple bond	Benzene ring
H—C—C—H (with H's)	H—C=C—H (with H's)	H—C≡C—H	benzene ring
Ethane	Ethene (ethylene)	Ethyne (acetylene)	Benzene

B. Combustion *(This may be an instructor demonstration.)*

When a compound burns in the presence of oxygen, the reaction is called combustion. This is the reaction that occurs when the methane gas in a Bunsen burner, a gas range, or a heater is ignited. The products of combustion are carbon dioxide (CO_2) and water (H_2O).

$$CH_4 + 2O_2(g) \xrightarrow{\text{Heat}} CO_2(g) + 2H_2O(g)$$

Methane

C. Bromine Test

When bromine (Br_2) reacts with an alkene, the dark red color of the Br_2 disappears quickly as the atoms of bromine bond with the carbon atoms in the double bond. If the red color disappears rapidly, we know the compound contains an unsaturated site.

$$
\underset{\textit{Colorless}}{CH_3-CH=CH-CH_3} \;+\; \underset{\textit{Red}}{Br_2} \longrightarrow \underset{\textit{Colorless}}{CH_3-\overset{\overset{\displaystyle Br}{|}}{CH}-\overset{\overset{\displaystyle Br}{|}}{CH}-CH_3}
$$

Bromine reacts with alkanes by replacing an H with a Br. However, the reaction is slow and requires light. Then the red bromine color persists for several minutes before it fades. Aromatic compounds (benzene ring) are not reactive with bromine. However, the methyl in toluene can react, but slowly.

$$
\underset{\textit{Colorless}}{CH_3CH_3} \;+\; \underset{\textit{Red}}{Br_2} \;\underset{\text{(slowly)}}{\overset{\text{Light}}{\longrightarrow}}\; \underset{\textit{Colorless}}{CH_3CH_2-Br} \;+\; \underset{\textit{Pungent odor}}{HBr(g)}
$$

$+\; Br_2 \longrightarrow$ No reaction

D. Potassium Permanganate ($KMnO_4$) Test

In this test, potassium permanganate ($KMnO_4$) reacts with alkenes, but not with alkanes or aromatic compounds. In the reaction, the purple color of $KMnO_4$ changes to the muddy brown of manganese dioxide (MnO_2). The product is a diol.

$$
\text{Alkene} \;+\; \underset{\textit{Purple}}{KMnO_4} \longrightarrow \text{Diol} \;+\; \underset{\textit{Brown solid}}{MnO_2(s)}
$$

Lab Information

Time: 2 hr
Comments: Tear out the report sheets and place them beside the matching procedures.
                   Caution: Hydrocarbons are flammable. Use very small amounts. Do not use any burners during these labs. Avoid touching the chemicals. Dispose of organic wastes in proper containers.

Related topics: Saturated, unsaturated, and aromatic hydrocarbons, haloalkanes, combustion, addition reactions

Experimental Procedures

Wear your safety goggles!

A. Types of Hydrocarbons

Materials: Organic model kit

Using the organic model kits, make models of ethene (ethylene), propene, cyclobutene, cis-2-butene, and ethyne (acetylene). Use the springs in the kit to form double bonds, triple bonds, and rings. A cis isomer of an alkene has the carbon groups attached on the same side of the double bond; a trans isomer has the groups attached on opposite sides. Draw their condensed structural formulas.

B. Combustion *(This may be an instructor demonstration.)*

Materials: Evaporating dish, wooden splints, matches, cyclohexane, cyclohexene, toluene, unknowns

B.1 *Working in the hood*, place 5 drops of cyclohexane on an evaporating dish. Using a lighted splint, *carefully* ignite the sample. Repeat the combustion test with 5 drops of cyclohexene, toluene, unknowns. Observe the flame and type of smoke produced by each. Record your observations.

B.2 Write the equations for the combustion reactions of cyclohexane (C_6H_{12}), cyclohexene (C_6H_{10}), and toluene (C_7H_8).

C. Bromine Test *(This may be an instructor demonstration.)*

Materials: 4 test tubes, test tube rack, dropper bottle of 1% bromine solution (in methylene chloride), cyclohexane, cyclohexene, toluene, unknowns

Caution: Work in the hood. The fumes of Br_2 can irritate the throat and sinuses. If bromine is spilled on the skin, flood with water for 10 minutes.

C.1 Place 15 drops of each hydrocarbon in a separate dry test tube. Label. Carefully add 3–4 drops of the bromine solution to each. Observe whether the red color disappears immediately or not. Hold the test tubes containing cyclohexane and toluene in a window with direct light. Observe whether the red color disappears, and if the odor of HBr given off can be detected.

C.2 Draw the condensed structural formula of each reactant, and of its products if a reaction occurred. If no reaction occurs, write "no reaction" (NR).

D. Potassium Permanganate ($KMnO_4$) Test

Materials: 4 test tubes, test tube rack, 1% $KMnO_4$, cyclohexane, cyclohexene, toluene, unknowns

Place 5 drops of each hydrocarbon in a separate test tube. Add 15 drops of 1% $KMnO_4$ solution. *Caution: $KMnO_4$ stains the skin.* A positive test for an unsaturated compound is a change in color from purple to brown in 60 seconds or less. Record your observations.

E. Identification of Unknown

From your test results, identify your unknown as a saturated (alkane) or unsaturated (alkene) hydrocarbon. Give your reasoning. If the names of the possible compounds are known, write their names and condensed structural formulas.

Report Sheet - Lab 23

Date _____   Name _____

Section _____   Team _____

Instructor _____

Pre-Lab Study Questions

1. What changes in color occur when bromine or $KMnO_4$ reacts with an alkene?

2. What are the products of combustion of an organic compound?

3. Why is the reaction of ethene with bromine called an addition reaction?

A. Types of Hydrocarbons

Models of Unsaturated Hydrocarbons

Name	Condensed Structural Formula
Ethene	
Propene	
Cyclobutene	
cis-2-Butene	
Ethyne (acetylene)	

Report Sheet - Lab 23

Questions and Problems

Q.1 Write the names of the following compounds:

a. CH_3—$CH=CH_2$ _____

b. $CH_3CH=CH$—CH_3 _____

c. $CH_2=C$—$CH_2CH_2CH_3$ with CH_3 branch _____

d. _____

e. $HC\equiv CH$ _____

f. _____

Q.2 Draw the structural feature that is characteristic of the following types of hydrocarbons:

Alkane	Alkene	Alkyne	Aromatic

Report Sheet - Lab 23

B. Combustion

Hydrocarbon	B.1 Observations of Combustion	B.2 Balanced Equation for Combustion
Cyclohexane (C_6H_{12})		
Cyclohexene (C_6H_{10})		
Toluene (C_7H_8)		
Unknown		

C. Bromine Test and D. Potassium Permanganate ($KMnO_4$) Test

	C.1 Bromine Test Observations	D. $KMnO_4$ Test Observations
Cyclohexane (C_6H_{12})		
Cyclohexene (C_6H_{10})		
Toluene (C_7H_8)		
Unknown		

Report Sheet - Lab 23

C.2

	Cyclohexane	Cyclohexene	Toluene
Condensed structural formula			
Product with bromine (if reaction)			

Questions and Problems

Q.3 Complete and balance the following reactions:

a. CH_3—CH=CH—CH_3 + Br_2 $\longrightarrow$

b. 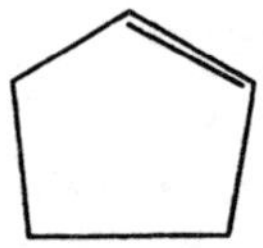 + Cl_2 $\longrightarrow$

E. Identification of Unknown

Results of tests with unknown

Unknown	Combustion	Bromine Test	KMnO$_4$ Test	Alkane or Alkene?

Explain your conclusion.

Goals

- Determine chemical and physical properties of alcohols and phenols.
- Classify an alcohol as primary, secondary, or tertiary.
- Perform a chemical test to distinguish between the classes of alcohols.
- Write the formulas of the oxidation products of alcohols.

Discussion

A. Structures of Alcohols and Phenol

Alcohols are organic compounds that contain the hydroxyl group (–OH). The simplest alcohol is methanol. Ethanol is found in alcoholic beverages and preservatives, and is used as a solvent. 2-Propanol, also known as rubbing alcohol, is found in astringents and perfumes.

CH_3OH

Methanol
(methyl alcohol)

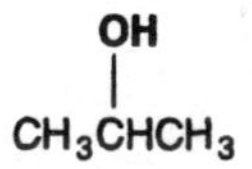

OH
|
CH_3CHCH_3

2-Propanol
(isopropyl alcohol)

CH_3CH_2OH

Ethanol
(ethyl alcohol)

A benzene ring with a hydroxyl group is known as phenol. Concentrated solutions of phenol are caustic and cause burns. However, derivatives of phenol, such as thymol, are used as antiseptics and are sometimes found in cough drops.

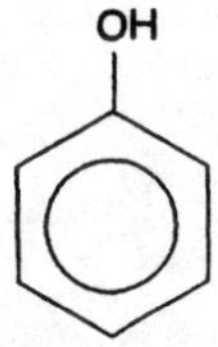

Phenol

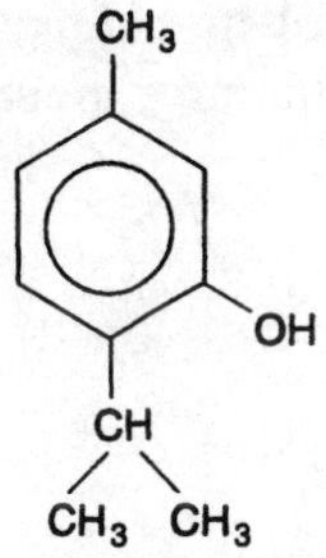

Thymol
(2-isopropyl-5-methylphenol)

Classification of Alcohols

In a primary (1°) alcohol, the carbon atom attached to the –OH group is bonded to one other carbon atom. In a secondary (2°) alcohol, it is attached to two carbon atoms and in a tertiary (3°) alcohol to three carbon atoms.

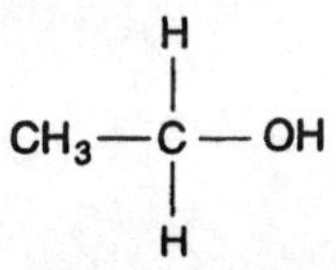

Ethanol
primary (1°) alcohol

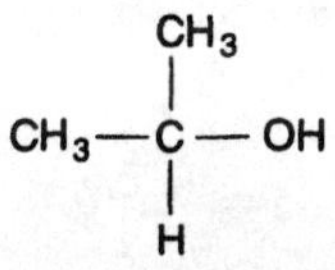

2-Propanol
secondary (2°) alcohol

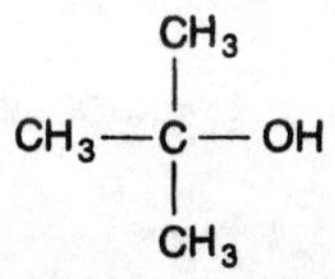

2-Methyl-2-Propanol
tertiary (3°) alcohol

B. Properties of Alcohols and Phenol

The polarity of the hydroxyl group (–OH) makes alcohols with four or fewer carbon atoms soluble in water because they can form hydrogen bonds. However, in longer-chain alcohols, a large hydrocarbon section makes them insoluble in water.

$$CH_3 - O - H$$

δ– δ+   Hydrogen bonding between the
hydroxyl group of methanol and water

δ– δ+
O — H
|
H

Acidity of Phenol

In water, phenol acts as a weak acid because the hydroxyl group ionizes slightly. Although phenol has six carbon atoms, its acid behavior makes it soluble in water.

Phenol   Phenoxide ion

C. Oxidation of Alcohols

Primary and secondary alcohols are easily oxidized. An oxidation consists of removing an H from the –OH group and another H from the C atom attached to the –OH group. Tertiary alcohols do not undergo oxidation because there are no H atoms on that C atom. Primary and secondary alcohols can be distinguished from tertiary alcohols using a solution with chromate, CrO_4^{2-}. An oxidation has occurred when the orange color of the chromate solution turns green.

$$CH_3 - CH_2 - OH \ + \ CrO_4^{2-} \ \xrightarrow{H^+} \ CH_3 - \overset{\overset{\displaystyle O}{\|}}{C} - H \ + \ Cr^{3+}$$

1° Alcohol   Orange   Aldehyde   Green

$$CH_3 - \overset{\overset{\displaystyle CH_3}{|}}{CH} - OH \ + \ CrO_4^{2-} \ \xrightarrow{H^+} \ CH_3 - \overset{\overset{\displaystyle O}{\|}}{C} - CH_3 \ + \ Cr^{3+}$$

2° Alcohol   Orange   Ketone   Green

$$CH_3 - \overset{\overset{\displaystyle CH_3}{|}}{\underset{\underset{\displaystyle CH_3}{|}}{C}} - OH \ + \ CrO_4^{2-} \ \xrightarrow{H^+} \ \text{No reaction (stays orange)}$$

3° Alcohol   Orange

D. Ferric Chloride Test

Phenols react with the Fe^{3+} ion in a ferric chloride ($FeCl_3$) solution to give complex ions with strong colors from red to purple.

$$\text{Phenol} \quad + \quad Fe^{3+} \quad \rightarrow \quad Fe^{3+} \cdot \text{ phenol complex}$$
$$\textit{colorless} \qquad\quad \textit{yellow} \qquad\qquad\qquad\qquad \textit{purple}$$

E. Identification of Unknown

The group of tests for alcohols and phenols described in this experiment will be used to identify the functional group and family of an unknown substance.

Lab Information

Time: 2 hr
Comments: Be careful when you work with chromate solution. It contains concentrated acid.
                Do not use burners in lab when you work with flammable organic compounds.
                Tear out the Lab report sheets and place them beside the matching procedures.
Related topics: Alcohols, classification of alcohols, solubility of alcohols in water, phenols, oxidation of alcohols

Experimental Procedures

GOGGLES MUST BE WORN!

A. Structures of Alcohols and Phenol

Materials: Organic model kits

Observe the models or obtain an organic model kit and construct models of ethanol, 2-propanol, and *t*-butyl alcohol (2-methyl-2-propanol). Write the condensed structural formula of each. Write the condensed structural formula for phenol. Classify each alcohol as a primary, secondary, or tertiary alcohol.

B. Properties of Alcohols and Phenol

Materials: 6 test tubes, pH paper, stirring rod, ethanol, 2-propanol, t-butyl alcohol (2-methyl-2-propanol), cyclohexanol, 20% phenol, and unknown

Odor Place 5 drops of each of the alcohols, phenol, and unknown to six separate test tubes. *Avoid skin contact with phenol.* Carefully detect the odor of each. Hold your breath as you gently fan some fumes from the top of the test tube toward you.

Solubility in water Add about 2 mL of water (40 drops) to each test tube. Shake and determine whether each alcohol is soluble or not. If the substance is soluble in water, you will see a clear solution with no separate layers. If it is insoluble, a cloudy mixture or separate layer will form. Record your observations.

Acidity Obtain a container of pH paper. Place a stirring rod in one of the alcohols and touch a drop to the pH paper. Compare the color of the paper with the chart on the container to determine the pH of the solution. Record.

DISPOSE OF ORGANIC SUBSTANCES IN DESIGNATED WASTE CONTAINERS!

C. Oxidation of Alcohols

Materials: 6 test tubes, ethanol, 2-propanol, *t*-butyl alcohol (2-methyl-2-propanol), cyclohexanol, phenol, unknown, 2% chromate solution

C.1 Place 8 drops of the alcohols and an unknown in separate test tubes. Carefully add 2 drops of chromate solution to each. Look for a color change in the chromate solution as you add it to the sample. If the orange color turns to green in 1–2 minutes, oxidation of the alcohol has taken place. If the color remains orange, no reaction has occurred. If a test tube becomes hot, place it in a beaker of ice-cold water. Record your observations. Caution: **Chromate solution contains concentrated H_2SO_4, which is corrosive.**

C.2 Draw the condensed structural formula of each alcohol.

C.3 Classify each alcohol·as primary (1°), secondary (2°), or tertiary (3°).

C.4 Draw the condensed structural formulas of the products where oxidation occurred. When there is no change in color, no oxidation took place. Write "no reaction" (NR).

D. Ferric Chloride Test

Materials: 6 test tubes, ethanol, 2-propanol, *t*-butyl alcohol (2-methyl-2-propanol), cyclohexanol, 20% phenol, unknown, 1% $FeCl_3$ solution

Place 5 drops of the alcohols and unknown in separate test tubes. Add 5 drops of 1% $FeCl_3$ solution to each. Stir and record observations.

DISPOSE OF ORGANIC SUBSTANCES IN DESIGNATED WASTE CONTAINERS!

E. Identification of Unknown Substance

Use the test results to identify your unknown as one of the five compounds used in this experiment.

Report Sheet - Lab 24

Date _____   Name _____

Section _____   Team _____

Instructor _____

Pre-Lab Study Questions

1. What is the functional group of an alcohol and a phenol?

2. Why are some alcohols soluble in water?

3. How are alcohols classified?

A. Structures of Alcohols and Phenols

Ethanol	2-Propanol
Classification:	
t-Butyl alcohol (2-methyl-2-propanol)	Phenol
Classification:	

Report Sheet - Lab 24

Questions and Problems

Q.1   Write the structures and classifications of the following alcohols:

1-Pentanol	3-Pentanol
Cyclopentanol	1-Methylcyclopentanol

Q.2   In your textbook or a chemistry handbook, look up and draw the condensed structural formulas and uses of thymol, menthol, and resorcinol. Circle the phenol functional group in each structure. If the compounds are available in lab, carefully note and describe their odors.

B.   Properties of Alcohols and Phenols

Alcohol	Odor	Soluble in Water?	pH
Ethanol			
2-propanol			
t-butyl alcohol			
Cyclohexanol			
Phenol			
Unknown			

Report Sheet - Lab 24

C. Oxidation of Alcohols

Alcohol	C.1 Color Change with CrO_4^{2-}	C.2 Condensed Structural Formula	C.3 Classification	C.4 Oxidation Product (If reaction takes place)
Ethanol				
2-Propanol				
t-butyl alcohol				
Cyclohexanol				
Phenol				
Unknown				

Questions and Problems

Q.3 Write the product of the following reactions (if no reaction, write NR):

a. $CH_3CH_2CH_2OH$ $\xrightarrow{[O]}$

b. $CH_3\overset{OH}{\underset{|}{C}}HCH_2CH_3$ $\xrightarrow{[O]}$

c. 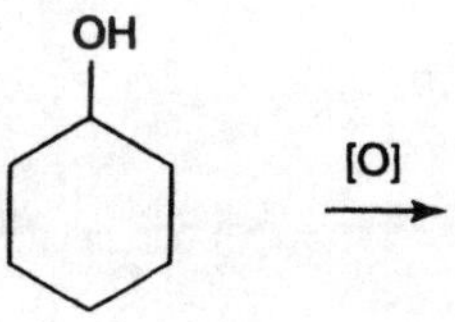 (cyclohexanol structure) $\xrightarrow{[O]}$

Report Sheet - Lab 24

D. Ferric Chloride Test

Alcohol	FeCl$_3$ Test
Ethanol	
2-propanol	
t-Butyl alcohol	
Cyclohexanol	
Phenol	
Unknown	

E. Identification of Unknown Substance

Unknown # _____

Summary of Testing	Results	Conclusions
B. Odor		
B. Soluble in water?		
B. pH		
C. Oxidation: CrO$_4^{2-}$		
D. FeCl$_3$		
Name of Unknown	Structure	

Aldehydes and Ketones

Goals

- Write the functional groups of aldehydes and ketones.
- Determine chemical and physical properties of aldehydes and ketones.
- Perform chemical tests to distinguish between aldehydes and ketones.

Discussion

A. Structures of Some Aldehydes and Ketones

Aldehydes and ketones both contain the carbonyl group. In an aldehyde, the carbonyl group has a hydrogen atom attached; the aldehyde functional group occurs at the end of the carbon chain. In a ketone, the carbonyl group is located between two of the carbon atoms within the chain.

Aldehydes   Ketone

$$-\overset{\overset{\displaystyle O}{\|}}{C}-$$ $$CH_3-\overset{\overset{\displaystyle O}{\|}}{C}-H$$ $$CH_3CH_2-\overset{\overset{\displaystyle O}{\|}}{C}-H$$ $$CH_3-\overset{\overset{\displaystyle O}{\|}}{C}-CH_3$$

Carbonyl   Acetaldehyde   Propionaldehyde   Acetone
functional group   (ethanal)   (propanal)   (2-propanone)

B. Properties of Aldehydes and Ketones

Many aldehydes and ketones have sharp odors. If you have taken a biology class, you may have noticed the odor of Formalin™, which is a solution of formaldehyde. When you remove fingernail polish, you may notice the strong odor of acetone, the simplest ketone, which is used as the solvent. Aromatic aldehydes have a variety of odors. Benzaldehyde, the simplest aromatic aldehyde, has an odor of almonds.

$$H-\overset{\overset{\displaystyle O}{\|}}{C}-H$$

Formaldehyde   Benzaldehyde

C. Iodoform Test for Methyl Ketones

Ketones containing a methyl group attached to the carbonyl give a reaction with iodine (I_2) in a NaOH solution. The reaction produces solid, yellow iodoform, CHI_3. Iodoform, which has a strong medicinal odor, is used as an antiseptic.

$$CH_3-\overset{\overset{\displaystyle O}{\|}}{C}-CH_3 \; + \; 3I_2 \; + \; 4NaOH \longrightarrow CH_3-\overset{\overset{\displaystyle O}{\|}}{C}-O^-Na^+ \; + \; CHI_3 \; + \; 3NaI \; + \; 3H_2O$$

Methyl ketone   Iodine   Iodoform
               (red)   (yellow)

D. Oxidation of Aldehydes and Ketones

Aldehydes are oxidized using Benedict's solution, which contains cupric ion, Cu^{2+}. Because ketones cannot oxidize, this test can distinguish aldehydes from ketones. In the oxidation reaction, the blue-green Cu^{2+} is reduced to cuprous ion (Cu^+), which forms a reddish-orange precipitate of Cu_2O.

$$CH_3-\overset{\overset{\displaystyle O}{\|}}{C}-H \quad + \quad 2Cu^{2+} \quad \longrightarrow \quad CH_3-\overset{\overset{\displaystyle O}{\|}}{C}-OH \quad + \quad Cu_2O(s)$$

Aldehyde   Blue   Red-orange

$$CH_3-\overset{\overset{\displaystyle O}{\|}}{C}-CH_3 \quad + \quad 2Cu^{2+} \quad \longrightarrow \quad \text{No reaction (stays blue)}$$

Ketone   Blue

E. Identification of an Unknown

Using the results of the tests, an unknown substance can be identified as an aldehyde or ketone.

Lab Information

Time: 2 hr

Comments: Flammable compounds are used in this experiment. Do not use burners.
In tests with color changes, carefully observe the color of the reactants before they are mixed.
Tear out the Lab report sheets and place them beside the matching procedures.

Related topics: Aldehydes, ketones, oxidation of aldehydes

Experimental Procedures

BE SURE TO WEAR YOUR GOGGLES!

A. Structures of Some Aldehydes and Ketones

Materials: Organic model kits

Make models or observe models of formaldehyde, acetaldehyde, propionaldehyde, acetone, butanone, and cyclohexanone. Draw their condensed structural formulas. Write the IUPAC and common names (if any) for each.

B. Properties of Aldehydes and Ketones

Materials: Chemistry handbook, test tubes, droppers, 5- or 10-mL graduated cylinder, acetone, benzaldehyde, camphor, vanillin, cinnamaldehyde, 2,3-butanedione, propionaldehyde, cyclohexanone, and unknown

Odors of Aldehydes and Ketones

B.1 Carefully detect the odor of samples of acetone, benzaldehyde, camphor, vanillin, cinnamaldehyde, and 2,3-butanedione, and unknown.

B.2 Draw their condensed structural formulas. You may need a chemistry handbook or a *Merck Index*. Identify each as a ketone or aldehyde.

Solubility of Aldehydes and Ketones

B.3 Place 2 mL of water in each of 4 separate test tubes. Add 5 drops of propionaldehyde (propanal), benzaldehyde, acetone, cyclohexanone, and unknown. Record your observations. *Save the samples for part C.*

C. Iodoform Test for Methyl Ketones

Materials: Test tubes from part B.3, dropper, 10% NaOH, warm water bath, and iodine test reagent

Using the test tubes from part B.3, add 10 drops of 10% NaOH to each. Warm the tubes in a warm water bath to 50–60°C. Add 20 drops of iodine test reagent. Look for the formation of a yellow solid precipitate. Record your results.

D. Oxidation of Aldehydes and Ketones

Materials: Test tubes, propionaldehyde (propanal), benzaldehyde, acetone, cyclohexanone, unknown, benedict's reagent, droppers, boiling water bath

Place 10 drops of propionaldehyde (propanal), benzaldehyde, acetone, cyclohexanone and unknown in separate test tubes. Label. Add 2 mL of Benedict's reagent to each test tube. Place the test tubes in a boiling water bath for 5 minutes. The appearance of the red-orange color of Cu_2O indicates that oxidation has occurred. Moderate amounts of Cu_2O will blend with the blue Cu^{2+} solution to form green or rust color. Record your observations. Identify the compounds that gave an oxidation reaction.

E. Identification of an Unknown

If you were given an unknown compound, you can now compare the results of the tests for the unknown with the tests you performed with known aldehydes and ketones. Identify your unknown as an aldehyde or a ketone.

Report Sheet - Lab 25

Date _____   Name _____

Section _____   Team _____

Instructor _____

Pre-Lab Study Questions

1. What is the functional group of an aldehyde? A ketone?

2. What is the oxidation product of an aldehyde?

A. Structures of Some Aldehydes and Ketones

Formaldehyde IUPAC Name _____	Acetaldehyde IUPAC Name_____
Propionaldehyde IUPAC Name_____	Acetone IUPAC Name_____
Butanone Common Name_____	Cyclohexanone

Report Sheet - Lab 25

B. Properties of Aldehydes and Ketones

	B.1 Odor	B.2 Condensed Structural Formula	Aldehyde or Ketone?
Acetone			
Benzaldehyde			
Camphor			
Vanillin			
Cinnamaldehyde			
2,3-butanedione			
Unknown			

Report Sheet - Lab 25

Questions and Problems

Q.1 What aldehyde or ketone might be present in the following everyday products?

Artificial butter flavor in popcorn   _____

Almond-flavored cookies   _____

Candies with cinnamon flavor   _____

Nail polish remover   _____

B., C., and D. Solubility, Iodoform, and Oxidation of Aldehydes and Ketones

	B.3 Solubility Soluble in water?	C. Iodoform Test Methyl ketone present?	D. Benedict's Test Oxidation occurred?
Propionaldehyde			
Benzaldehyde			
Acetone			
Cyclohexanone			
Unknown			

Questions and Problems

Q.2 Complete the following with the word *soluble or insoluble:*

Aldehydes and ketones containing one to four carbon atoms are _____ in water.

Aldehydes and ketones containing five or more carbon atoms are _____ in water.

Report Sheet - Lab 25

Q.3 Indicate the test results for each of the following compounds in the iodoform test and in the Benedict's test:

	Iodoform Test	Benedict's Test
$\underset{\displaystyle CH_3CCH_2CH_3}{\overset{\displaystyle O \atop \displaystyle \|}{}}$		
$\underset{\displaystyle CH_3CH}{\overset{\displaystyle O \atop \displaystyle \|}{}}$		
$\underset{\displaystyle CH_3CH_2CCH_2CH_3}{\overset{\displaystyle O \atop \displaystyle \|}{}}$		
$\underset{\displaystyle CH_3CCH_2CH}{\overset{\displaystyle O \quad O \atop \displaystyle \| \quad \|}{}}$		

Q.4 Two compounds, A and B, have the formula of C_3H_6O. Determine their condensed structural formulas and names using the following test results.

a. Compound A forms a red-orange precipitate with Benedict's reagent but does not react with iodoform.

b. Compound B forms a yellow solid in the iodoform test but does not react with Benedict's reagent.

Report Sheet - Lab 25

E. Identification of an Unknown Substance
Unknown _____

Summary of Testing	Results	Conclusion
Odor		Does it have a familiar odor?
Solubility in water		How many carbon atoms?
Iodoform test		Is it a methyl ketone?
Benedict's test		Is it an aldehyde?

Questions and Problems

Q.5 What can you conclude about the structure and functional group of your unknown? Explain.

Q.6 What chemical tests could you use to distinguish between 2-pentanone and 3-pentanone?

Report Sheet - Lab 26

Date _____   Name _____

Section _____   Team _____

Instructor _____

Pre-Lab Study Questions

1. What are some sources of carbohydrates in your diet?

2. What does the D in D-glucose mean?

3. What is the bond that links monosaccharides in di- and polysaccharides?

A. Monosaccharides

A.1 Fischer projections

L-glyceraldehyde   D-glyceraldehyde

How does L-glyceraldehyde differ from D-glyceraldehyde?

A.2 Fischer projection of D-glucose   Haworth (cyclic) formulas

α-D-glucose β-D-glucose

Report Sheet - Lab 26

A.3 Fischer projection of D-fructose   Haworth (cyclic) formula for α-D-fructose

Fischer projection of D-galactose   Haworth (cyclic) formula for α-D-galactose

Questions and Problems

Q.1 How does the structure of D-glucose compare to the structure of D-galactose?

B. Disaccharides

B.1 Structure of α-D-maltose

B.2 Equation for the hydrolysis of α-D-maltose

Report Sheet - Lab 26

B.3 Formation of α-D-lactose

B.4 Structure of sucrose

Questions and Problems

Q.2 What is the type of glycosidic bond in maltose?

Q.3 Why does maltose have both α and β anomers? Explain.

C. Polysaccharides

C.1 A portion of amylose

C.2 Comparison of amylopectin to amylose

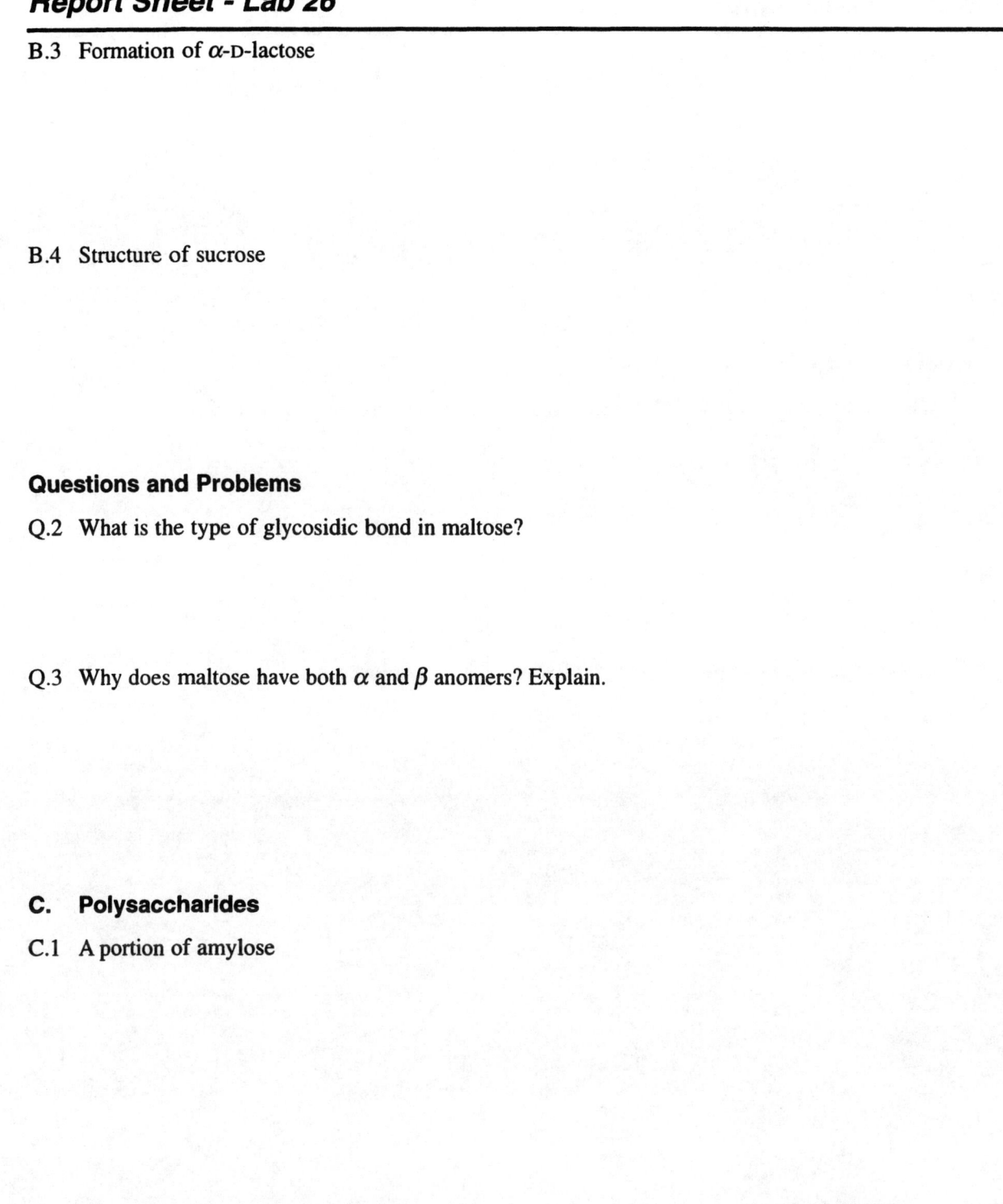

Report Sheet - Lab 26

C.3 A portion of cellulose

Questions and Problems

Q.4 What is the monosaccharide that results from the complete hydrolysis of amylose?

Q.5 What is the difference in the structure of amylose and cellulose?

Goals

- Observe physical and chemical properties of some common carbohydrates.
- Use physical and chemical tests to distinguish between monosaccharides, disaccharides, and polysaccharides.
- Identify an unknown carbohydrate.
- Relate the process of digestion to the hydrolysis of carbohydrates.

Discussion

A. Benedict's Test for Reducing Sugars

All of the monosaccharides and most of the disaccharides can be oxidized. When the cyclic structure opens, the aldehyde group is available for oxidation. Benedict's reagent contains Cu^{2+} ion that is reduced. Therefore, all the sugars that react with Benedict's reagent are called *reducing sugars*. Ketoses also act as reducing sugars because the ketone group on carbon 2 isomerizes to give an aldehyde group on carbon 1.

When oxidation of a sugar occurs, the Cu^{2+} is reduced to Cu^+, which forms a red precipitate of cuprous oxide, $Cu_2O(s)$. The color of the precipitate varies from green to gold to red depending on the concentration of the reducing sugar.

Sucrose is not a reducing sugar because it cannot revert to the open-chain form that would provide the aldehyde group needed to reduce the cupric ion.

Sucrose

B. Seliwanoff's Test for Ketoses

Seliwanoff's test is used to distinguish between hexoses with a ketone group and hexoses that are aldehydes. With ketoses, a deep red color is formed rapidly. Aldoses give a light pink color that takes a longer time to develop. The test is most sensitive for fructose, which is a ketose.

C. Fermentation Test

Most monosaccharides and disaccharides undergo fermentation in the presence of yeast. The products of fermentation are ethyl alcohol (CH_3CH_2OH) and carbon dioxide (CO_2). The formation of bubbles of carbon dioxide is used to confirm the fermentation process.

$$C_6H_{12}O_6 \xrightarrow{\text{yeast}} 2C_2H_5OH + 2CO_2(g)$$
$$\text{Glucose} \qquad\qquad\quad \text{Ethanol}$$

Although enzymes are present for the hydrolysis of most disaccharides, they are not available for lactose. The enzymes needed for the fermentation of galactose are not present in yeast. Lactose and galactose give negative results with the fermentation test.

D. Iodine Test for Polysaccharides

When iodine (I_2) is added to amylose, the helical shape of the unbranched polysaccharide traps iodine molecules, producing a deep blue-black complex. Amylopectin, cellulose, and glycogen react with iodine to give red to brown colors. Glycogen produces a reddish-purple color. Monosaccharides and disaccharides are too small to trap iodine molecules and do not form dark colors with iodine.

E. Hydrolysis of Disaccharides and Polysaccharides

Disaccharides hydrolyze in the presence of an acid to give the individual monosaccharides.

$$\text{Sucrose} + H_2O \xrightarrow{H^+} \text{Glucose} + \text{Fructose}$$

In the laboratory, we use water and acid to hydrolyze starches, which produce smaller saccharides such as maltose. Eventually, the hydrolysis reaction converts maltose to glucose molecules. In the body, enzymes in our saliva and from the pancreas carry out the hydrolysis. Complete hydrolysis produces glucose, which provides about 50% of our nutritional calories.

$$\text{Amylose, amylopectin} \xrightarrow[\text{amylase}]{H^+ \text{ or}} \text{dextrins} \xrightarrow[\text{amylase}]{H^+ \text{ or}} \text{maltose} \xrightarrow[\text{maltase}]{H^+ \text{ or}} \text{many D-glucose units}$$

F. Testing Foods for Carbohydrates

Several of the tests such as the iodine test can be carried out with food products such as cereals, bread, crackers, and pasta. Some of the carbohydrates we have discussed can be identified.

Lab Information

Time: 3 hr
Comments: Tear out the report sheets and place them next to the matching procedures. .
Related Topics: Carbohydrates, hemiacetals, aldohexoses, ketohexoses, reducing sugars, fermentation

Experimental Procedures

A. Benedict's Test for Reducing Sugars

Materials: Test tubes, 400-mL beaker, droppers, hot plate or Bunsen burner, 5- or 10-mL graduated cylinder, Benedict's reagent, 2% carbohydrate solutions: glucose, fructose, sucrose, lactose, starch, and an unknown

Place 10 drops of solutions of glucose, fructose, sucrose, lactose, starch, water, and unknown in separate test tubes. Label each test tube. Add 2 mL of Benedict's reagent to each sample. Place the test tubes in a boiling water bath for 3–4 minutes. The formation of a greenish to reddish-orange color indicates the presence of a reducing sugar. If the solution is the same color as the Benedict's reagent in water (the control), there has been no oxidation reaction. Record your observations. Classify each as a reducing or nonreducing sugar.

B. Seliwanoff's Test for Ketoses

Materials: Test tubes, 400-mL beaker, droppers, hot plate or Bunsen burner, 5- or 10-mL graduated cylinder, Seliwanoff's reagent, 2% carbohydrate solutions: glucose, fructose, sucrose, lactose, starch, and an unknown

Place 10 drops of solutions of glucose, fructose, sucrose, lactose, starch, water, and unknown in separate test tubes. Add 2 mL of Seliwanoff's reagent to each. *The reagent contains concentrated HCl. Use carefully.*

Place the test tubes in a boiling hot water bath and note the time. After 1 minute, observe the colors in the test tubes. A rapid formation of a deep red color indicates the presence of a ketose. Record your results as a fast color change, slow change, or no change.

C. Fermentation Test

Materials: Fermentation tubes (or small and large test tubes), baker's yeast, 2% carbohydrate solutions: glucose, fructose, sucrose, lactose, starch, and an unknown

Fill fermentation tubes with a solution of glucose, fructose, sucrose, lactose, starch, water, and unknown. Add 0.2 g of yeast to each and mix well. See Figure 27.1.

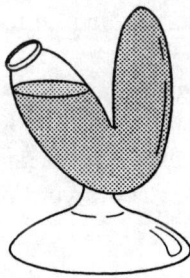

Figure 27.1   Fermentation tube filled with a carbohydrate solution

If fermentation tubes are not available, use small test tubes placed upside down in larger test tubes. Cover the mouth of the large test tube with filter paper or cardboard. Place your hand firmly over the paper cover and invert. When the small test tube inside has completely filled with the mixture, return the larger test tube to an upright position. See Figure 27.2.

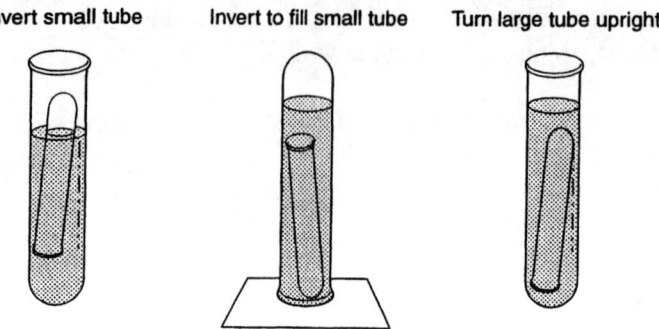

Invert small tube   Invert to fill small tube   Turn large tube upright

Figure 27.2   Test tubes used as fermentation tubes

Set the tubes aside. At the end of the laboratory period, and again at the next laboratory period, look for gas bubbles in the fermentation tubes or inside the small tubes. Record your observations. See Figure 27.3.

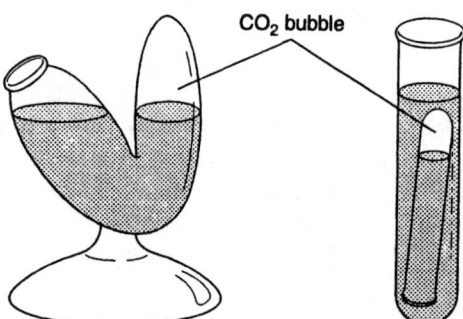

CO_2 bubble

Figure 27.3   Fermentation tubes with CO_2 bubbles

D.   Iodine Test for Polysaccharides

Materials: Spot plate or test tubes, droppers, iodine reagent, 2% carbohydrate solutions in dropper bottles: glucose, fructose, sucrose, lactose, starch, and an unknown

Using a spot plate, place 5 drops of each solution of glucose, fructose, sucrose, lactose, starch, water, and unknown in the wells. (If you do not have a spot plate, use small test tubes.) Add 1 drop of iodine solution to each sample. A dark blue-black color is a positive test for amylose in starch. A red or brown color indicates the presence of other polysaccharides. Record your results. Complete the table to identify your unknown.

E.   Hydrolysis of Disaccharides and Polysaccharides

Materials: Test tubes, 10-mL graduated cylinder, 400-mL beaker (boiling water bath), hot plate or Bunsen burner, spot plate or watch glass, 10% HCl, 10% NaOH, red litmus paper, iodine reagent, Benedict's reagent, 2% starch and sucrose solutions in dropper bottles

Place 3 mL of 2% starch in two test tubes and 3 mL of 2% sucrose solution in two more test tubes. To one sample each of sucrose and starch, add 20 drops of 10% HCl. To the other samples of sucrose and starch, add 20 drops of H_2O. Label the test tubes and heat in a boiling water bath for 10 minutes.

Remove the test tubes from the water bath and let them cool. To the samples containing HCl, add 10% NaOH (about 20 drops) until one drop of the mixture turns litmus paper blue, indicating the HCl has been neutralized. Test the samples for hydrolysis as follows:

Iodine Test Place 5 drops of each solution on a spot plate or watch glass. Add 1 drop of iodine reagent to each. Record observations. Determine if hydrolysis has occurred in each.

Benedict's Test Add 2 mL of Benedict's reagent to each of the samples and heat in a boiling water bath for 3–4 minutes. Determine if hydrolysis has occurred in each.

F.   Testing Foods for Carbohydrates

Materials: Sugar samples (refined, brown, "natural," powdered), honey, syrups (corn, maple, fruit), foods with starches: cereals, pasta, bread, crackers, potato, Benedict's solution, Seliwanoff's reagent, iodine reagent

Obtain two carbohydrate samples to test. Perform the Benedict's, Seliwanoff's, and iodine tests on each. Describe the kinds of carbohydrates you identify in each sample.

Report Sheet - Lab 27

Date _____   Name _____

Section _____   Team _____

Instructor _____

Pre-Lab Study Questions

1. What happens to glucose or galactose when the Cu^{2+} in Benedict's is reduced?

2. Would you expect fructose or glucose to form a red color rapidly with Seliwanoff's reagent?

3. Why don't all the disaccharides undergo fermentation with yeast?

4. How can the iodine test be used to distinguish between amylose and glycogen?

Results of Carbohydrate Tests

	A. Benedict's Test	B. Seliwanoff's Test	C. Fermentation Test	D. Iodine Test
Glucose				
Fructose				
Sucrose				
Lactose				
Starch				
Water				
Unknown #_____				

Report Sheet - Lab 27

Questions and Problems

Q.1 From the results in part A, list the sugars that are reducing sugars and those that are not.

Reducing sugars

Nonreducing sugars

Q.2 What sugars are ketoses?

Q.3 What sugars give a positive fermentation test?

Q.4 Which carbohydrates give a blue-black color in the iodine test?

Identifying an Unknown Carbohydrate

Unknown No._____

	Results with Unknown	Possible Sugars Present
Benedict's (A)		
Seliwanoff's (B)		
Fermentation (C)		
Iodine (D)		

What carbohydrate(s) is/are in your unknown?

Report Sheet - Lab 27

Questions and Problems

Q.5 What carbohydrate(s) would have the following test results?
   a. Produces a reddish-orange solid with Benedict's and a red color with Seliwanoff's reagent in 1 minute

   b. Gives a color change with Benedict's test, a light orange color with Seliwanoff's reagent after 5 minutes, and produces no bubbles during fermentation

   c. Gives no color change with Benedict's or Seliwanoff's test, but turns a blue-black color with iodine reagent

E.   Hydrolysis of Disaccharides and Polysaccharides

Results	Sucrose + H_2O	Sucrose + HCl	Starch + H_2O	Starch + HCl
Iodine test				
Benedict's test				
Hydrolysis products present				

Questions and Problems

Q.6 How do the results of the Benedict's test indicate that hydrolysis of sucrose and starch occurred?

Q.7 How do the results of the iodine test indicate that hydrolysis of starch occurred?

Report Sheet - Lab 27

Q.8   Indicate whether the following carbohydrates will give a positive (+) or a negative (-) result in each type of test listed below:

	Benedict's Test	Seliwanoff's Test	Fermentation Test	Iodine Test
Glucose				
Fructose				
Galactose				
Sucrose				
Lactose				
Maltose				
Amylose				
Amylopectin				

F.   Testing Foods for Carbohydrates

	Food Item 1	Food Item 2
Benedict's test		
Seliwanoff's test		
Iodine test		
Possible carbohydrates present		

Carboxylic Acids and Esters

Goals

- Write the structural formulas of carboxylic acids and esters.
- Determine the solubility and acidity of carboxylic acids and their salts.
- Write equations for neutralization and esterification of acids.
- Prepare esters and identify their characteristic odors.

Discussion

A. Carboxylic Acids and Their Salts

A salad dressing made of oil and vinegar tastes tart because it contains vinegar, which is known as acetic acid (ethanoic acid). The sour taste of fruits such as lemons is due to acids such as citric acid. Face creams contain alpha hydroxy acids such as glycolic acid. All these acids are carboxylic acids, which contain the carboxyl group: a carbonyl group attached to a hydroxyl group. A dicarboxylic acid such as malonic acid, found in apples, has two carboxylic acid functional groups. The carboxylic acid of benzene is called benzoic acid.

Ionization of Carboxylic Acids in Water

Carboxylic acids are weak acids because the carboxylic acid group ionizes slightly in water to give a proton and a carboxylate ion. However, like the alcohols, the polarity of the carboxylic acid group makes acids with one to four carbon atoms soluble in water. Acids with two or more carboxyl groups (diacids) are more soluble in water.

Neutralization of Carboxylic Acids

An important feature of carboxylic acids is their neutralization by bases such as sodium hydroxide to form carboxylate salts and water. We saw in an earlier experiment that neutralization is the reaction of an acid with a base to give a salt and water. Even insoluble carboxylic acids with five or more carbon atoms can be neutralized to give corresponding salts that are usually soluble in water. For this reason, acids used in food products or medications are in their soluble salt form rather than the acid itself.

$$HX \quad + \quad NaOH \quad \longrightarrow \quad Na^+X^- \quad + \quad H_2O$$

Acid   Base   Salt   Water

$$CH_3-\overset{\overset{\displaystyle O}{\|}}{C}-OH \quad + \quad NaOH \quad \longrightarrow \quad CH_3-\overset{\overset{\displaystyle O}{\|}}{C}-O^-Na^+ \quad + \quad H_2O$$

Acetic acid   Sodium acetate
(a carboxylate salt)

B. Esters

Carboxylic acids may have tart or unpleasant odors, but many esters have pleasant flavors and fragrant odors. Octyl acetate gives oranges their characteristic odor and flavor; pear flavor is due to pentyl acetate. The flavor and odor of raspberries come from isobutyl formate.

$$CH_3(CH_2)_7O-\overset{\overset{\displaystyle O}{\|}}{C}-CH_3 \qquad CH_3(CH_2)_4-O-\overset{\overset{\displaystyle O}{\|}}{C}-CH_3 \qquad CH_3-\overset{\overset{\displaystyle CH_3}{|}}{C}HCH_2-O-\overset{\overset{\displaystyle O}{\|}}{C}-H$$

Octyl acetate   Pentyl acetate   Isobutyl formate
(oranges)   (pears)   (raspberries)

An ester of salicylic acid is methyl salicylate, which gives the flavor and odor of oil of wintergreen used in candies and ointments for sore muscles. When salicylic acid reacts with acetic anhydride, acetylsalicylic acid (ASA) is formed, which is aspirin, widely used to reduce fever and inflammation.

Methyl salicylate
(wintergreen)

Acetylsalicylic acid
(aspirin)

Esterification and Hydrolysis

In a reaction called *esterification,* the carboxylic acid group combines with the hydroxyl group of an

Esterification ⟶

$$CH_3-\overset{\overset{\displaystyle O}{\|}}{C}-OH \; + \; HO-(CH_2)_4-CH_3 \underset{}{\overset{H^+}{\rightleftharpoons}} CH_3-\overset{\overset{\displaystyle O}{\|}}{C}-O-(CH_2)_4-CH_3 \; + \; H_2O$$

Acetic acid   1-Pentanol   Pentyl acetate (pear flavor)

⟵ **Hydrolysis**

alcohol. The reaction, which takes place in the presence of an acid, produces an ester and water. The reverse reaction, hydrolysis, occurs when an acid catalyst and water cause the decomposition of an ester to yield the carboxylic acid and alcohol. The ester product is favored when an excess of acid or alcohol is used; hydrolysis is favored when more water is used.

C. Hydrolysis of Esters

When an ester is hydrolyzed in the presence of a base, the reaction is called saponification. The products are the salt of the carboxylic acid and the alcohol. Although the ester is usually insoluble in water, the salt and alcohol (if short-chain) are soluble.

Ester bond splits

$$CH_3-\overset{\overset{\textstyle O}{\|}}{C}-O-CH_2CH_3 \; + \; NaOH \; \longrightarrow \; CH_3-\overset{\overset{\textstyle O}{\|}}{C}-O^-\,Na^+ \; + \; HO-CH_2CH_3$$

Ethyl acetate   Sodium acetate   Ethanol
(ester)   (carboxylate salt)   (alcohol)

Lab Information

Time: 2 hr
Comments: When noting odors, hold your breath and fan across the top of a test tube to detect the odor. The formation of esters requires concentrated acid. Use carefully.
                 Tear out the report sheets and place them beside the matching procedure.
Related Topics: Carboxylic acid functional group, ester functional group, ionization of carboxylic acids, neutralization, esterification, hydrolysis, saponification

Experimental Procedures

BE SURE TO WEAR YOUR SAFETY GOGGLES!

A. Carboxylic Acids and Their Salts

Materials: Test tubes, glacial acetic acid, benzoic acid(s), dropper, spatula, pH paper, red and blue litmus paper, stirring rod, 400-mL beaker, hot plate or Bunsen burner, 10% NaOH, 10% HCl

A.1 Write the structural formulas for acetic acid and benzoic acid.

A.2 Place about 2 mL of water in two test tubes. Add 5 drops of acetic acid to one test tube and a small amount of benzoic solid (enough to cover the tip of a spatula) to the other. Tap the sides of the test tubes to mix or stir with a stirring rod. Identify the acid that dissolves.

A.3 Test the pH of each carboxylic acid by dipping a stirring rod into the solution, then touching it to a piece of pH paper. Compare the color on the paper with the color chart on the container and report the pH.

A.4 Place the test tube of benzoic acid (solid should be present) in a hot water bath and heat for 5 minutes. Describe the effect of heating on the solubility of acid. Allow the test tube to cool. Record your observations.

A.5 Add about 10 drops of NaOH to each test tube until a drop of the solution turns red litmus paper blue. Record your observations. Write the equations for the reactions with NaOH including the structures of the sodium salts formed.

A.6 Add about 10 drops of HCl to each sample until it is neutralized (blue litmus paper turns red). Record your observations. Write equations for the reactions of the sodium salts that formed.

B. Esters

Materials: Organic model set, test tubes, hot plate or Bunsen burner, 400-mL beaker, stirring rod, spatula, small beaker, methanol, 1-pentanol, 1-octanol, benzyl alcohol, 1-propanol, salicylic acid(s), glacial acetic acid, H_3PO_4

B.1   Make a model of acetic acid and methyl alcohol. Remove the components of water and form an ester bond to give methyl acetate. Write the equation for the formation of the ester.

B.2   As assigned, prepare one of the mixtures listed by placing 3 mL of the alcohol in a test tube and label it with the mixture number. Add 2 mL of a carboxylic acid or the amount of solid that covers the tip of a spatula. Write the condensed structural formulas for the alcohols and carboxylic acids.

Mixture	Alcohol	Carboxylic Acid
1	Methanol	Salicylic acid
2	1-Pentanol	Acetic acid
3	1-Octanol	Acetic acid
4	Benzyl alcohol	Acetic acid
5	1-Propanol	Acetic acid

Caution: Use care in dispensing glacial acetic acid. It can cause burns and blisters on the skin.

B.3   With the test tube pointed away from you, *cautiously* add 15 drops of concentrated phosphoric acid, H_3PO_4. Stir. Place the test tube in a boiling water bath for 15 minutes. Remove the test tube and *cautiously* fan the vapors toward you. Record the odors you detect such as pear, banana, orange, raspberry, or oil of wintergreen (spearmint). For a stronger odor, place 15 mL of hot water in a small beaker and pour the ester solution into the hot water.

If specified by your instructor, repeat the esterification with other mixtures of alcohol and carboxylic acid. Note the odors of esters produced by other students. Write the condensed structural formulas and names of the esters produced. *Dispose of the ester products as instructed*.

C. Hydrolysis of Esters

Materials: Test tube, test tube holder, droppers, stirring rod, hot plate or Bunsen burner, 250- or 400-mL beaker, methyl salicylate in a dropper bottle, 10% NaOH, 10% HCl, blue litmus paper

C.1   Draw the condensed structural formula of methyl salicylate.

C.2   Place 3 mL of water in a test tube. Add 5 drops of methyl salicylate. Record the appearance and odor of the ester.

C.3   Add 1 mL (20 drops) of 10% NaOH. There should be two layers in the test tube. Place the test tube in a boiling water bath for 30 minutes or until the top layer of the ester disappears. Record any changes in the odor of the ester. Remove the test tube and cool in cold water.

C.4   Write the equation for the saponification reaction.

C.5   After the solution is cool, add about 20 drops (1 mL) of 10% HCl until a drop of the solution turns blue litmus paper red. Record your observations. Determine the formula of the solid that forms.

Report Sheet - Lab 28

Date _____   Name _____

Section _____   Team _____

Instructor _____

Pre-Lab Study Questions

1. What are some carboxylic acids you encounter in daily life?

2. What are esters?

A.   Carboxylic Acids and Their Salts

	Acetic Acid	Benzoic Acid
A.1 Condensed structural formulas		
A.2 Solubility in cold water		
A.3 pH		
A.4 Solubility in hot water	✕	
A.5 NaOH		
A.6 HCl		

Write equations for the reactions of the carboxylic acids with NaOH (A.5).

Write equations for the reactions of the salts of the carboxylic acids with HCl (A.6).

Report Sheet - Lab 28

Questions and Problems

Q.1   How does NaOH affect the solubility of benzoic acid in water? Why?

Q.2   Write the names of the following carboxylic acids and esters:

a.
$$CH_3CH_2CH_2—\overset{\overset{\displaystyle O}{\|}}{C}—OH$$

b.
$$CH_3CH_2\overset{\overset{\displaystyle O}{\|}}{C}—OCH_3$$

Q.3   Why are there differences in the solubility of the carboxylic acids in part A?

B.   Esters

B.1   Equation for the formation of methyl acetate:

Report Sheet - Lab 28

B.2 Condensed Structural Formulas of Alcohol and Carboxylic Acid	B.3 Odor of Ester	Condensed Structural Formula and Name of Ester
Methanol and salicylic acid		
1-Pentanol and acetic acid		
1-Octanol and acetic acid		
Benzyl alcohol and acetic acid		
1-Propanol and acetic acid		

Report Sheet - Lab 28

C. Hydrolysis of Esters

C.1 Condensed structural formula of methyl salicylate	
C.2 Appearance and odor of methyl salicylate	
C.3 Describe the appearance and odor of the ester after adding NaOH and heating	
C.4 Write the equation for the saponification of the ester	
C.5 What changes occur when HCl is added?	
What is the formula of the compound that formed when HCl was added?	

Aspirin and Other Analgesics

Goals

- Use an esterification reaction to synthesize aspirin.
- Purify the crude aspirin sample.
- Test the purity of prepared aspirin and commercial aspirin products.
- Determine the physical and chemical properties of aspirin.
- Use thin-layer chromatography to separate and identify substances in analgesics.

Discussion

In the 18th century, an extract of willow bark was found useful in reducing fevers (antipyretic) and relieving pain and inflammation. Although salicylic acid was effective at reducing fever and pain, it damaged the mucous membranes of the mouth and esophagus, and caused hemorrhaging of the stomach lining. At the turn of the century, scientists at the Bayer Company in Germany noted that salicylic acid contained a phenol group that might cause the damage. They decided to modify the salicylic acid by forming an ester with a two-carbon acetyl group. The resulting substance was acetylsalicylic acid, or ASA, which we call aspirin. Aspirin acts by inhibiting the formation of prostaglandins, 20-carbon acids that form at the site of an injury and cause inflammation and pain.

In commercial aspirin products, a small amount of acetylsalicylic acid (300 mg to 400 mg) is bound together with a starch binder and sometimes caffeine and buffers to make an aspirin tablet. The basic conditions in the small intestine break down the acetylsalicylic acid to yield salicylic acid, which is absorbed into the bloodstream. The addition of a buffer reduces the irritation caused by the carboxylic acid group of the aspirin molecule.

A. Preparation of Aspirin

Aspirin (acetylsalicylic acid) can be prepared from acetic acid and the hydroxyl group on salicylic acid. However, this is a slow reaction. The ester forms rapidly when acetic anhydride is used to provide the acetyl group. *The aspirin you will prepare in this experiment is impure and must not be taken internally!*

Using the following equation, the maximum amount (yield) of aspirin that is possible from 2.00 g of salicylic acid can be calculated.

$$2.00 \text{ g salicylic acid} \times \frac{1 \text{ mole salicylic acid}}{138 \text{ g}} \times \frac{1 \text{ mole aspirin}}{1 \text{ mole salicylic acid}} \times \frac{180 \text{ g}}{1 \text{ mole aspirin}}$$

$$= 2.61 \text{ g aspirin (possible)}$$

Suppose the total amount of aspirin you obtain has a mass of 2.25 g. A percentage yield can be calculated as follows:

$$\% \text{ Yield} = \frac{\text{g aspirin obtained}}{\text{g aspirin calculated}} \times 100 = 86.2\% \text{ yield of aspirin product}$$

B. Testing Aspirin Products

The purity of the crude sample and the recrystallized aspirin product can be tested with ferric chloride, $FeCl_3$. The Fe^{3+} ion reacts with the phenol group on salicylic acid and gives a purple color. This test can also be used to determine the purity of commercially prepared aspirin. Sometimes old aspirin breaks down to give salicylic acid and acetic acid. Then the aspirin in the bottle smells like vinegar and should be discarded.

C. Analysis of Analgesics

Aspirin is one of several analgesics that are used to relieve pain. Other analgesics include acetaminophen, ibuprofen, and naproxen. Many aspirin products include caffeine. These products including aspirin are used to reduce fever, which means they are also antipyretics. However, aspirin also has anti-inflammatory properties and may reduce the risk of a heart attack.

Aspirin   **Acetaminophen**   **Ibuprofen**

Naproxen   **Caffeine**

Thin-Layer Chromatography (TLC)

Thin-layer chromatography (TLC) is a technique used to separate substances in a mixture. A TLC plate is typically a sheet of plastic, coated with a thin layer of a solid adsorbent such as silica gel.

Small amounts of known and unknown substances are placed as small spots at one end of the TLC plate. Then the end of the plate with the spots is placed in a solvent contained in a developing chamber.

The solid silica layer on the TLC plate is called the *stationary phase*. The solvent or the *moving phase* slowly moves up the silica layer on the TLC plate carrying the substances in the spot with it. The more soluble a substance is in the solvent, the higher the solvent will carry it up the plate. A substance that adheres strongly to the stationary silica gel moves only a short distance with the solvent. Thus, differences in the substances determines the distances they travel up the plate.

As the solvent front nears the top of the TLC plate, the plate is removed, marked, and dried. Then the substances are visualized. If they have colors, they can be seen directly. In this experiment, they are colorless. Because the silica material on the plate contains a fluorescent compound, ultraviolet light (254-nm) from a UV lamp can be used to visualize the substances, which appear as dark spots on the plate.

Calculating R_f Values

A value called the R_f value can be calculated for each substance on a plate. The R_f is the distance that a substance moves on the plate divided by the distance the solvent moves. An unknown substance is identified if its R_f value matches the R_f value of one of the known substances used on the plate. (See Figure 29.1.)

$$R_f = \frac{\text{distance substance moves}}{\text{distance solvent moves}}$$

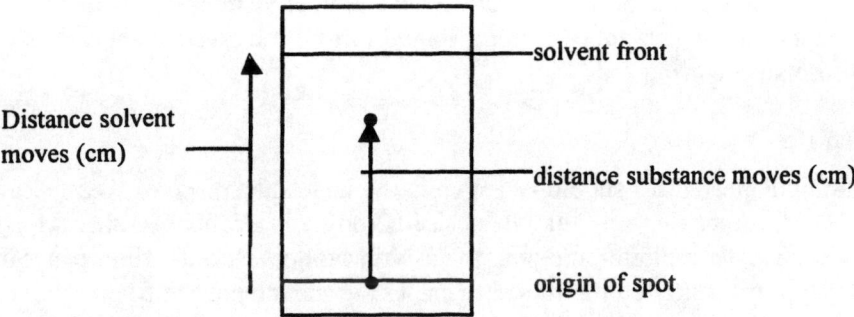

Figure 29.1 Distances moved by a substance and solvent on a TLC plate

In this experiment, you will use TLC to determine the R_f values for several known analgesics. You will also determine the R_f values and identify the types of analgesics in a variety of over the counter drugs used to relieve pain.

Experimental Procedures

WEAR PROTECTIVE GOGGLES AT ALL TIMES!

A. Preparation of Aspirin

Materials: 125-mL Erlenmeyer flask, 400-mL beaker, hot plate or Bunsen burner, ice, salicylic acid, acetic anhydride, 5- or 10-mL graduated cylinder, stirring rod, pan or large beaker, dropper, 85% H_3PO_4 in a dropper bottle, Büchner filtration apparatus, filter paper, spatula, watch glass

A.1 Weigh a 125-mL Erlenmeyer flask. Add 2.00 g of salicylic acid and reweigh. *Working in the hood, carefully* add 5 mL of acetic anhydride to the flask.

Caution: Acetic anhydride is irritating to the nose and sinus. Handle carefully.

Slowly add 10 drops of 85% phosphoric acid, H_3PO_4. Stir the mixture with a stirring rod. Place the flask and its contents in a boiling water bath and stir until all the solid dissolves.

Remove the flask from the hot water and let it cool to room temperature. *Working in the hood, cautiously* add 20 drops of water to the cooled mixture.

> **KEEP YOUR FACE AWAY FROM THE TOP OF THE FLASK: ACETIC ACID VAPORS ARE IRRITATING.**

When the reaction is complete, add 50 mL of cold water. Cool the mixture by placing the flask in an ice bath for 10 minutes. Stir. Crystals of aspirin should form. If no crystals appear, gently scratch the sides of the flask with a stirring rod.

Collecting the Aspirin Crystals

Some Büchner filtration apparatuses should be set up in the lab. Add a piece of filter paper. Place the funnel in the filter flask making sure that the neck fits snugly in a rubber washer. Moisten the filter paper. Turn on the water aspirator and pour the aspirin product onto the filter paper in the Büchner funnel. Push down gently on the funnel to create the suction needed to pull the water off the aspirin product. The aspirin crystals will collect on the filter paper. See Figure 29.2.

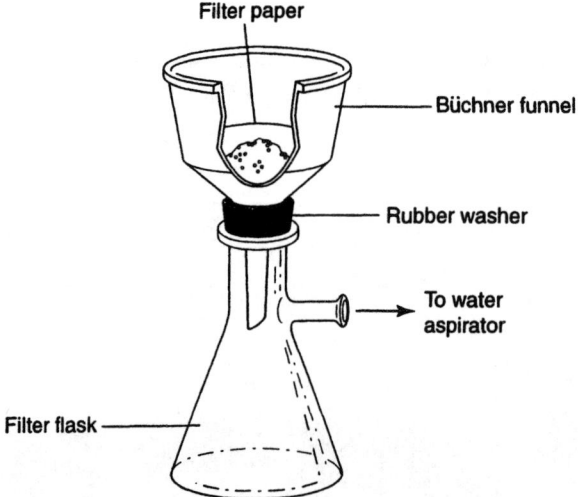

Figure 29.2 Apparatus for suction filtration with a Büchner funnel

Use a spatula to transfer any crystals left in the flask. Rinse the inside of the flask with a 10-mL portion of ice cold water to transfer all the crystals to the funnel. Wash the aspirin crystals on the filter paper with two 10-mL portions of cold water.

Spread the aspirin crystals out on the filter paper and draw air through the funnel. This helps dry the crystals. Turn off the water and use a spatula to lift and transfer the filter paper and aspirin to a paper towel. Don't touch it; it may still contain acid. Allow the crystals to air dry.

A.2 Weigh a clean, dry watch glass. Transfer the aspirin crystals to the watch glass and reweigh.

Calculations

A.3 Calculate the mass of salicylic acid.

A.4 Calculate the maximum yield of aspirin possible from the salicylic acid.

A.5 Calculate the mass of the aspirin you collected.

A.6 Calculate the percent yield of aspirin.

A.7 If a melting point apparatus is available, determine the melting point of your aspirin product. Pure aspirin has a melting point of 135°C. Salicylic acid melts at 157–159°C. Compare the melting point of your aspirin with the known melting points of aspirin and salicylic acid.

B. Testing Aspirin Products

Materials: Test tubes, spatula, aspirin from part A, commercial aspirin tablets, buffered aspirin, acetylsalicylic acid, 0.15% (m/v) salicylic acid, pH indicator paper, stirring rod, 1% $FeCl_3$, 10% NaOH, 10% HCl, 400-mL beaker, hot plate or Bunsen burner, blue litmus paper

B.1 **pH of aspirin** Place 3 mL of 0.15% salicylic acid in the first test tube. In test tubes, 2–5, place a few crystals (the amount on the tip of a spatula) of the following substances and add 3 mL of water to each:

1.	0.15% Salicylic acid	4.	Aspirin product from part A
2.	Commercial aspirin (crushed)	5.	Acetylsalicylic acid
3.	Buffered aspirin (crushed)		

Stir each mixture and touch the stirring rod to a piece of pH indicator paper. Compare the color of the paper to the chart on the container. Record the pH of each. *Save these test tubes and samples for part B.2.*

B.2 **Testing aspirin purity** To each of the samples from B.1, add 5 drops of 1% ferric chloride ($FeCl_3$) solution. Any free salicylic acid (unreacted during synthesis or resulting from hydrolysis in the breakdown of aspirin) reacts with the $FeCl_3$ to give a purple color. The more salicylic acid in the sample, the deeper the color. This indicates that the product is impure or that decomposition has taken place.

The maximum salicylic acid allowed in commercially prepared aspirin products is 0.15%. If the sample test has a lighter color than a 0.15% standard, the sample would be considered *pure* by USP standards. If the sample is darker, it is *impure* and not safe for ingestion. However, no matter what the results of the test, your laboratory-prepared aspirin must not be ingested. Record the colors. Compare the purity of the tested products to the reference sample of salicylic acid.

C. Analysis of Analgesics

Materials: 400-mL beaker (developing chamber), Saran wrap, rubber band to fit beaker, solvent (75% ethyl acetate and 25% hexane), TLC plate with silica gel, UV lamp (short wave 254 nm), micropipettes, spot plates, dropper bottle containing 1% solutions in ethanol of aspirin, ibuprofen, acetaminophen, naproxen, caffeine, over the counter drugs, ruler

Preparing the TLC Developing Chamber

Obtain a 400-mL beaker, a piece of Saran wrap that covers, and a rubber band. Carefully pour a small amount of solvent into the beaker to a level of 0.5 – 0.6 cm. *It is important that the solvent level is below the spots you place on the TLC plate.* Cover the beaker with Saran wrap and secure with a rubber band.

Spotting the TLC Plate

Obtain a TLC plate that is 6 cm × 10 cm. *Be sure you handle the plates at the edge only to avoid transferring substances from your fingers.* Draw a light line with pencil about 1 cm above the end. This is your starting line or *origin.* Mark 6 dots on the line equally spaced. Label the dots from 1 through 6.

Place a few drops of each of the 1% solutions in a spot plate. Number the wells as follows.
  1. aspirin 2. ibuprofen   3. acetaminophen
  4. naproxen   5. caffeine   6. over the counter drug

Using clean capillary pipettes, one for each substance, spot a tiny amount of each substance on a dot. Lightly tap the micropipette to deliver a small amount. When dry you can apply again. The spot must be kept small rather than allowed to flow into larger spots. (See Figure 29.3.)

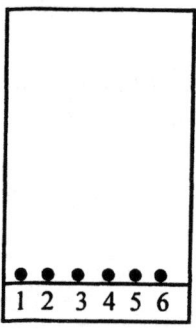

Figure 29.3   Spots of analgesics on a TLC plate

Placing TLC Plate in Developing Chamber

Carefully set the plate in the solvent in the beaker you prepared as the developing chamber. *The solvent must be lower than the origin on the plate.* Cover the beaker with Saran. Allow the beaker to remain undisturbed as the solvent moves up the TLC plate. (See Figure 29.4.)

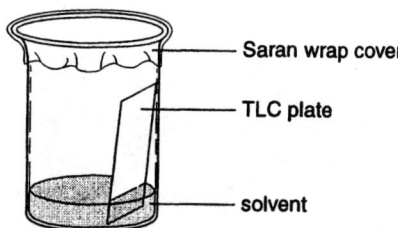

- Saran wrap cover
- TLC plate
- solvent

Figure 29.4   A developing chamber containing TLC plate.

When the solvent has risen almost to the top of the plate, open the chamber and draw a pencil line along the solvent front. Remove the plate and allow the solvent to evaporate in the hood. Place used solvent in the organic solvent container.

C.1   Observe the TLC place under UV light. Circle each spot. Draw a picture of the spots on your TLC plate.

C.2   Measure the distance from the origin to the solvent front. Measure the distance from the origin to the center of each spot.

C.3   Calculate the R_f value for each analgesic.

C.4   Identify the analgesics in the over the counter pain reliever.

Report Sheet - Lab 29

Date _____   Name _____

Section _____   Team _____

Instructor _____

Pre-Lab Study Questions

1. What functional group of aspirin causes it to irritate the stomach?

2. Why are buffers added to some aspirin products?

3. What quantity of aspirin is contained in most over-the-counter aspirin products?

Report Sheet - Lab 29

A. Preparation of Aspirin

A.1 Mass of flask

_____

Mass of flask and salicylic acid

_____

A.2 Mass of watch glass

_____

Mass of watch glass and crude aspirin product

_____

Calculations

A.3 Mass of salicyclic acid

_____

A.4 Possible (maximum) yield of aspirin

_____

(*Show calculations.*)

A.5 Mass of aspirin

_____

A.6 Percent yield

_____

(*Show calculations.*)

A.7 Melting point (°C) of aspirin product (optional)

_____

Questions and Problems

Q.1 Write the structural formula for aspirin. Label the ester group and the carboxylic acid group.

Q.2 If a typical aspirin tablet contains 325 mg aspirin (the rest is starch binder), how many tablets could you prepare from the aspirin you made in lab?

Report Sheet - Lab 29

B. Testing Aspirin Products

Samples Tested	B.1 pH	B.2 Color with $FeCl_3$
1. 0.15% Salicylic acid		
2. Commercial aspirin brand:		
3. Buffered aspirin brand:		
4. Aspirin from A		
5. Acetylsalicylic acid		

Questions and Problems

Q.3 Give an explanation for any differences in the pH values in the samples you tested in part B.1.

Q.4 How does the pH of buffered aspirin product compare to the pH of the nonbuffered aspirin product?

Q.5 What substance is present if the $FeCl_3$ test gives a purple color? Which sample is the most impure?

Q.6 Aspirin that has been stored for a long time may give a vinegar odor and give a purple color with $FeCl_3$. What reaction would cause this to happen?

Report Sheet - Lab 29

C. Analysis of Analgesics

C.1

C.2 Distance moved by solvent _____ cm

Spot #	Analgesic substance	C.2 Distance moved	C.3 R_f value
1			
2			
3			
4			
5			
6			

C.4 Substance(s) present in over the counter pain reliever
According to your R_f values, what substance(s) can you identify as present in the over the counter pain reliever in your analysis?

Substance 1 _____

Substance 2 _____

Goals

- Observe the physical and chemical properties of some common lipids.
- Draw the structure of a typical triacylglycerol.
- Distinguish between saturated and unsaturated fats.
- Determine the degree of unsaturation of some fats.
- Prepare a hand lotion and determine the function of its components.

Discussion

A. Triacylglycerols

The triacylglycerols, commonly called fats or oils, are esters of glycerol and fatty acids. Fatty acids are long-chain carboxylic acids, usually 14 to 18 carbons in length. When the fatty acid contains double bonds, the triacylglycerol is referred to as an unsaturated fat. When the fatty acid consists of an alkane-like carbon chain, the triacylglycerol is a saturated fat. Table 30.1 gives the formulas of the common fatty acids and their melting points. At room temperature, saturated fats are usually solid and unsaturated fats are usually liquid.

Table 30.1 *Formulas, Melting Points, and Sources of Some Fatty Acids*

Carbon Atoms	Structural Formula	Melting Point (°C)	Common Name	Source
Saturated fatty acids (single carbon–carbon bonds)				
12	$CH_3(CH_2)_{10}COOH$	44	lauric	coconut
14	$CH_3(CH_2)_{12}COOH$	54	myristic	nutmeg
16	$CH_3(CH_2)_{14}COOH$	63	palmitic	palm
18	$CH_3(CH_2)_{16}COOH$	70	stearic	animal fat
Monounsaturated fatty acids (one cis double bond)				
16	$CH_3(CH_2)_5CH=CH(CH_2)_7COOH$	1	palmitoleic	butter fat
18	$CH_3(CH_2)_7CH=CH(CH_2)_7COOH$	4	oleic	olives, corn
Polyunsaturated fatty acids (two or more cis double bonds)				
18	$CH_3(CH_2)_4CH=CHCH_2CH=CH(CH_2)_7COOH$	-5	linoleic	safflower, sunflower
18	$CH_3CH_2CH=CHCH_2CH=CHCH_2CH=CH(CH_2)_7COOH$	-11	linolenic	corn

Fats that contain mostly saturated fatty acids have a higher melting point than fats with more unsaturated fatty acids.

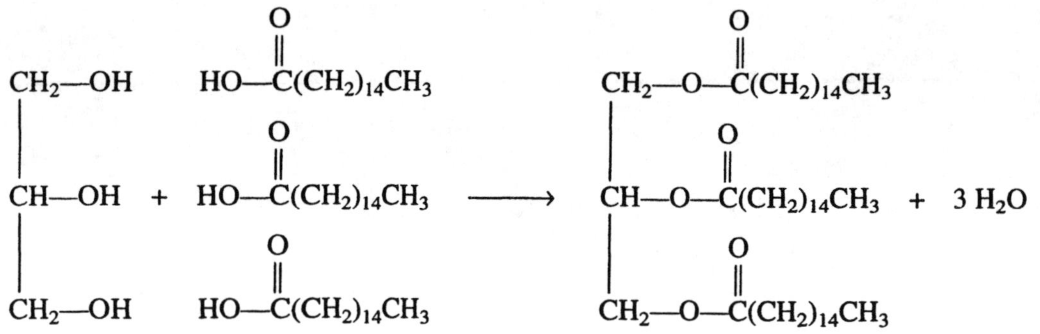

Glycerol   3 Palmitic acids   Glyceryl palmitate (tripalmitin)

B. Physical Properties of Some Lipids and Fatty Acids

Lipids are a family of compounds that are grouped by similarities in solubility rather than structure. As a group, lipids are more soluble in nonpolar solvents such as ether, chloroform, or benzene. Most are not soluble in water. Important types of lipids include fats and oils, glycerophospholipids, and steroids. Compounds classified as lipids include fat-soluble vitamins A, D, E, and K; cholesterol; hormones; portions of cell membranes; and vegetable oils. Table 30.2 lists the classes of lipids.

Table 30.2 *Classes of Lipid Molecules*

Lipids	Components
Waxes	Fatty acid and long-chain alcohol
Fats and oils (triacylglycerols)	Fatty acids and glycerol
Glycerophospholipids	Fatty acids, glycerol, phosphate, amino alcohol
Sphingolipids	Fatty acids, sphingosine, phosphate, amino alcohol
Glycosphinolipids	Fatty acids, glycerophospholipids sphingosine, monosaccharides
Steroids	A fused structure of three cyclohexanes and a cyclopentane

The structural formulas of three typical lipids are shown below:

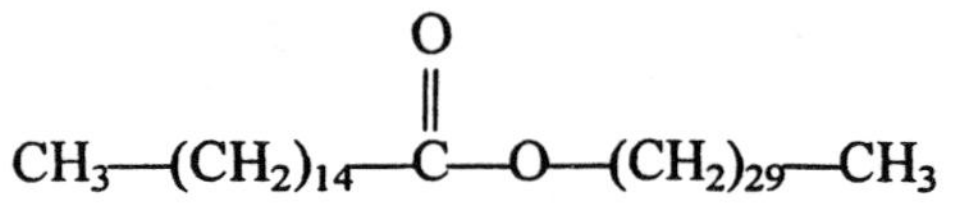

Wax

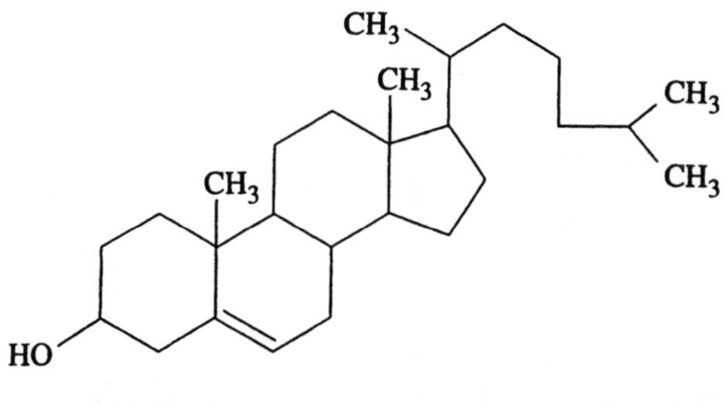

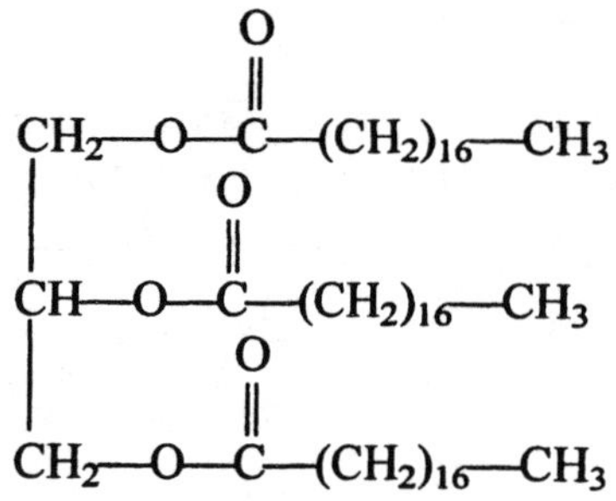

Triacylglycerol, a fat

Cholesterol, a steroid

C. Bromine Test for Unsaturation

The presence of unsaturation in a fatty acid or a triacylglycerol can be detected by the bromine test, which you used in an earlier experiment to detect double bonds in alkenes. If the orange color of the bromine solution fades quickly, an addition reaction has occurred and the oil or fat is unsaturated.

Bromine adds to the double bond

$$
\begin{array}{l}
CH_2-O-\overset{\overset{O}{\|}}{C}(CH_2)_7CH{=}CH(CH_2)_7CH_3 \\
| \qquad\quad \overset{O}{\|} \\
CH-O-\overset{\|}{C}(CH_2)_{16}CH_3 \qquad\qquad + Br_2 \longrightarrow \\
| \qquad\quad \overset{O}{\|} \\
CH_2-O-\overset{\|}{C}(CH_2)_{16}CH_3
\end{array}
$$

$$
\begin{array}{l}
CH_2-O-\overset{\overset{O}{\|}}{C}(CH_2)_7\overset{\overset{Br}{|}}{C}H{-}\overset{\overset{Br}{|}}{C}H(CH_2)_7CH_3 \\
| \qquad\quad \overset{O}{\|} \\
CH-O-\overset{\|}{C}(CH_2)_{16}CH_3 \\
| \qquad\quad \overset{O}{\|} \\
CH_2-O-\overset{\|}{C}(CH_2)_{16}CH_3
\end{array}
$$

D. Preparation of Hand Lotion

We use hand lotions and creams to soften our skin and reduce dryness. Typically, the formulation of a hand lotion consists of several components such as stearic acid, lanolin, triethanolamine, cetyl alcohol, glycerin (glycerol), water, and usually a fragrance. Lanolin from wool consists of a mixture of waxes.

Cetyl alcohol $\qquad$ $CH_3(CH_2)_{15}OH$

Stearic acid $\qquad$ $CH_3-(CH_2)_{16}-\overset{\overset{O}{\|}}{C}-OH$

Glycerol (glycerine) $\qquad$
$$
\begin{array}{l}
CH_2-OH \\
| \\
CH-OH \\
| \\
CH_2-OH
\end{array}
$$

Triethanolamine $\qquad$
$$
HOCH_2CH_2-\overset{\overset{\textstyle CH_2CH_2OH}{|}}{N}-CH_2CH_2OH
$$

Because lipids are nonpolar, they protect and soften by preventing the loss of moisture from the skin. Some of the components help emulsify the polar and nonpolar ingredients. In this experiment, we will see how the physical and chemical properties of lipids are used to prepare a hand lotion.

Lab Information

Time: $\qquad$ 3 hr

Comments: $\qquad$ Bromine can cause severe chemical burns. Use carefully.
Tear out the Lab report sheets and place them beside the matching procedures.

Related topics: Fatty acids, saturated and unsaturated fatty acids, lipids, triglycerides

Experimental Procedures

BE SURE YOU ARE WEARING YOUR SAFETY GOGGLES!

A. Triacylglycerols

Materials: Organic model kits or models

A.1 Use an organic model kit or observe prepared models of a molecule of glycerol and three molecules of ethanoic acid. What are the functional groups on each? Draw their structures.

A.2 Form ester bonds between the hydroxy groups on glycerol and the carboxylic acid groups of the ethanoic acid molecules. In the process, three molecules of water are removed. Write an equation for the formation of the glyceryl ethanoate.

Carry out the reverse process, which is hydrolysis. Add the components of water to break the ester bond. Add an arrow to the equation to show the reverse direction for the hydrolysis reaction.

B. Physical Properties of Some Lipids and Fatty Acids

Materials: Test tubes and stoppers, dropper bottles or solids: stearic acid, oleic acid, olive oil, safflower oil, lecithin, cholesterol, vitamin A capsules, spatulas, CH_2Cl_2 (optional)

To seven separate test tubes, add 5 drops or the amount of solid lipid held on the tip of a spatula: stearic acid, oleic acid, olive oil, safflower oil, lecithin, cholesterol, vitamin A (puncture a capsule or use cod liver oil).

Appearance and Odor

B.1 Classify each as a triacylglycerol (fat or oil), fatty acid, steroid, or phospholipid.

B.2 Describe their appearance.

B.3 Describe their odors.

Solubility in a Polar Solvent *(May be a demonstration)*

B.4 Add about 2 mL of water to each of the test tubes. Stopper and shake each test tube. Record your observations.

Solubility in a Nonpolar Solvent *(Optional or demonstration)*

B.5 Place 5 drops, or a small amount of solid, of the following in separate test tubes: stearic acid, oleic oil, olive oil, safflower oil, lecithin, cholesterol, and vitamin A. Add 1 mL (20 drops) methylene chloride, CH_2Cl_2, to each sample. Record the solubility of the lipids. *Save the test tubes and samples of stearic acid oleic acid, olive oil and safflower oil for part C.*

C. Bromine Test for Unsaturation

Materials: Samples from B.5, 1% Br_2 in methylene chloride

To the samples from B.5, add 1% bromine solution drop by drop until a permanent red-orange color is obtained or until 20 drops have been added.

Caution: Avoid contact with bromine solution; it can cause painful burns. Do not breathe the fumes.

Record your observations. Determine if the red-orange color fades rapidly or persists.

D. Preparation of Hand Lotion

Materials: Stearic acid, cetyl alcohol, lanolin (anhydrous), triethanolamine, glycerol, ethanol, distilled water, fragrance (optional), commercial hand lotions, 10-mL graduated cylinder, 50-mL graduated cylinder, 50-mL or 100-mL beakers, thermometer, 100-mL beakers, 250-mL beaker for water bath, Bunsen burner, iron ring, wire screen, stirring rods, tongs, pH paper

Team project: D.1, D.2, and D.3 may be prepared by different teams in the lab.

D.1 Obtain the following substances and combine in two 50-mL or 100-mL beakers. Use a laboratory balance to weigh out the solid substances. Use a 10-mL graduated cylinder to measure small volumes, and a 50-mL graduated cylinder to measure larger volumes.

Beaker 1		**Beaker 2**	
Stearic acid	3 g	Glycerin	2 mL
Cetyl alcohol	1 g	Water	50 mL
Lanolin (anhydrous)	2 g		
Triethanolamine	1 mL		

Water bath: Fill a 250-mL beaker about 2/3 full of water. Place the beaker on an iron ring covered with a wire screen. Lower a second iron ring that fits around the 250-mL beaker to stabilize it. Turn on the Bunsen burner to heat the water.

Using a pair of crucible tongs, hold Beaker 1 (four ingredients) in the water bath and heat to 80°C or until all the compounds have melted. Remove.

Using a pair of crucible tongs, place Beaker 2 (two ingredients) in the same water bath and heat to 80°C.

While still warm, slowly pour the glycerol-water mixture from Beaker 2 into Beaker 1 (four ingredients) as you stir. Add 5.0 mL of ethanol and a few drops of fragrance, if desired. Continue to stir for 3-5 minutes until a smooth, creamy lotion is obtained. If the resulting product is too thick, add more warm water.

Describe the smoothness and appearance of the hand lotion.

D.2 Repeat the experiment, but omit the triethanolamine from Beaker 1. Compare the properties of the resulting hand lotion to the one obtained in D.1 and to commercial hand lotions.

D.3 Repeat the experiment, but omit the stearic acid from Beaker 1. Compare the properties and textures of the resulting hand lotion to the one obtained in D.1 and to commercial hand lotions.

D.4 Determine the pH of the hand lotions you and others in your lab have prepared along with any commercial hand lotions.

Report Sheet - Lab 30

Date _____   Name _____

Section _____   Team _____

Instructor _____

Pre-Lab Study Questions

1. What is the functional group in a triacylglycerol?

2. Write the structure of linolenic acid. Why is it an unsaturated fatty acid?

A.   Triacylglycerols

A.1   Structure of glycerol   Structure of ethanoic acid

A.2   Reversible equation for the esterification and hydrolysis of glyceryl ethanoate

Report Sheet - Lab 30

B. Physical Properties of Some Lipids and Fatty Acids

Lipid	B.1 Type of Lipid	B.2 Appearance	B.3 Odor	B.4 Soluble in Water?	B.5 Soluble in CH_2Cl_2?
Stearic acid					
Oleic acid					
Olive oil					
Safflower oil					
Lecithin					
Cholesterol					
Vitamin A					

Questions and Problems

Q.1 Why are the compounds in part B classified as lipids?

Q.2 What type of solvent is needed to remove an oil spot? Why?

Report Sheet - Lab 30

C. Bromine Test for Unsaturation

Compound	Drops of Bromine Solution	Saturated or Unsaturated?
Fatty acids		
Stearic acid		
Oleic acid		
Triacylglycerols		
Safflower oil		
Olive oil		

Questions and Problems

Q.3 a. Write the condensed structural formulas of stearic acid and oleic acid.

Stearic acid

Oleic acid

b. Which fatty acid is unsaturated?

c. The melting point of stearic acid is 70°C, and that of oleic acid is 4°C. Explain the difference.

d. From the results of experiment C, how can you tell which is more unsaturated, oleic acid or stearic acid?

D. Preparation of Hand Lotion

Descriptions of the hand lotions		
D.1	D.2 (without triethanolamine)	D.3 (without stearic acid)
D.4 pH of the hand lotions		
pH of commercial hand lotions		
Brand_____	Brand_____	Brand_____
pH _____	pH _____	pH _____

Questions and Problems

Q. 4   How does omitting triethanolamine affect the properties and appearance of the hand lotion?

Q. 5   How does omitting stearic acid affect the properties and appearance of the hand lotion?

Q. 6   What would be a reason to have triethaonolamine and stearic acid as ingredients in hand lotion?

Glycerophospholipids and Steroids

Goals

- Observe the physical and chemical properties of some common lipids.

- Draw the structure of a typical glycerophospholipid.

- Draw the structure of cholesterol.

- Isolate lecithin and cholesterol from egg yolk.

Egg yolks contain about 10% lecithin, a glycerophospholipid, as well as cholesterol, a steroid. Other sources of lecithin are soybeans, peanuts, beef, and cauliflower. Lecithin contains two fatty acids (nonpolar) along with the ester of the amino alcohol choline (polar). The fatty acids found in lecithin are palmitic 11.7%, stearic 4.0%, palmitoleic 8.6%, oleic 9.8%, linoleic 55.0%, and linolenic 4.0%. Because lecithin consists of both polar and nonpolar sections, it is used to stabilize foods such as mayonnaise, salad dressings and Hollandaise sauce. Cholesterol is a type of lipid that has a steroid ring structure and used in the synthesis of several hormones

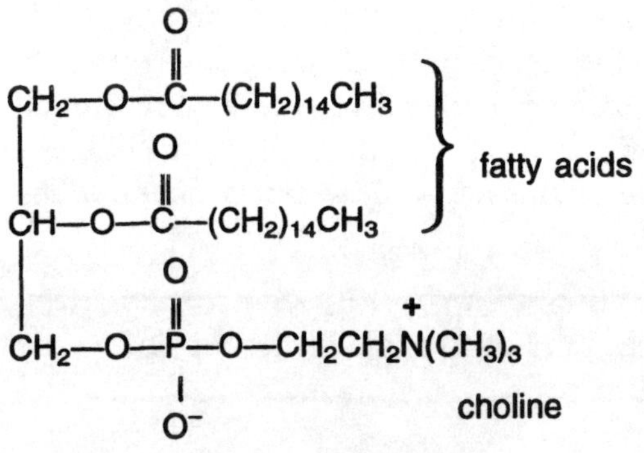

Lecithin, a glycerophospholipid

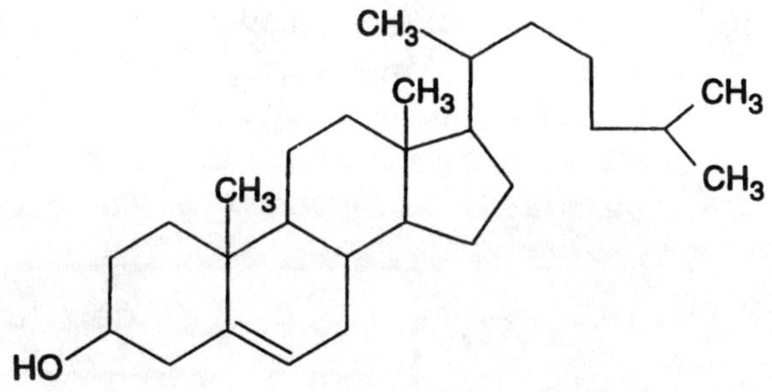

Cholesterol, a steroid

Lecithin and cholesterol are extracted from egg yolk using the organic solvents acetone and ethyl ether. Because cholesterol is soluble in acetone and lecithin is not, acetone is used to extract cholesterol from the egg yolk material. Then ethyl ether is used to extract lecithin from the remaining yolk material.

Lab Information

Time: 2-3 hr

Comments: Acetone and ethyl ether are *highly flammable*. Work with these solvents in the hood. No flames are allowed in lab during this experiment.
Tear out the Lab report sheets and place them beside the matching procedures.

Related topics: Fatty acids, glycerophospholipids, steroids

Experimental Procedures

BE SURE YOU ARE WEARING YOUR SAFETY GOGGLES!

A. Isolating Cholesterol from Egg Yolk

Materials: 1 hard boiled egg, 250-mL beaker, acetone, 125-mL Erlenmeyer flask, 50-mL flask, steam bath, 100-mL beaker, spatula, short-stem funnel, glass wool, stirring rods, melting point apparatus

> **Caution: No open flames are allowed in lab during this experiment.**
> **Organic solvents are flammable!**

A.1 Weigh a 250-mL beaker. Obtain an egg, peel it, and discard the egg white. Place the yolk in the beaker and break it up with a spatula or stirring rod. Determine the combined mass of the beaker and egg yolk.

Separating Cholesterol from Egg Yolk

Add 50 mL of acetone to the egg yolk in the 250-mL beaker. Stir the mixture periodically for the next 10-15 minutes. Allow the solid to settle to the bottom of the beaker. Slowly pour the acetone layer into a 125-mL Erlenmeyer flask. This acetone layer contains the cholesterol from the egg yolk. The solid material remaining in the beaker contains the lecithin. Place this beaker in a hood to allow the residual acetone to evaporate from the solid material. *Save this beaker and the solid material for the isolation of lecithin in part B.*

> **Carry out this part of the lab in the hood!**

A.2 Place a small amount of glass wool in a short-stem funnel. Pour the acetone layer containing the cholesterol in the 125-mL Erlenmeyer flask through the glass wool into a 100-ml beaker. Place the beaker on a steam bath to reduce the volume of the extract to between 5 to 10 mL. You can place 5-10 mL in a matching beaker and mark the level to determine the level for the acetone mixture. *Do not let it evaporate to dryness.*

Weigh a 50-mL Erlenmeyer flask. Pour the reduced volume of acetone extract containing cholesterol into the flask. Cork and place the flask in an ice bath for at least 20 minutes. A white solid, which is crude cholesterol, should form. *While you are waiting for the cholesterol crystals to appear, proceed with the isolation of lecithin in B.*

After white solid has formed, remove the flask containing the cholesterol from the ice bath and pour off the cold acetone. Place the acetone in a container marked "waste acetone".

A.3 Set the flask in the hood and allow residual acetone to evaporate. When dry, weigh the flask and cholesterol crystals. Record.

A.4 Describe the color and appearance of cholesterol.

A.5 (*optional*) Determine the melting point of your crude cholesterol. The known melting point for cholesterol is 149°C.

Calculations

A.6 Calculate the mass of the egg yolk.

A.7 Calculate the mass of the cholesterol.

A.8 Determine the percent by mass of cholesterol in egg yolk.

$$\text{Mass \% cholesterol} = \frac{\text{g cholesterol}}{\text{g egg yolk}} \times 100$$

B. Isolating Lecithin from Egg Yolk

Materials: Egg yolk residue from Part A, ethyl ether, hot plate or steam bath, 100-mL beaker, stirring rods

B.1 Place the beaker from Part A containing the egg yolk residue on a steam bath at a low or moderate temperature to evaporate any remaining acetone without splattering. Remove if it starts to splatter. There will be only a small amount of acetone in the residue, so this takes only a few minutes. Remove the beaker and let it cool.

> **Caution: Ether is *extremely* flammable.**
> **Carry out this part of the lab in the hood!**
> **No open flames are allowed during this experiment.**

In the hood, add 30 mL of ethyl ether to the egg yolk residue to extract the lecithin. Stir strongly for 2-3 minutes.

Weigh a clean 100-mL beaker and record. Pour the ether extract, which contains the lecithin, into the weighed 100-mL beaker. Place the left over egg yolk material in a "waste" container as indicated by your instructor.

B.2 *In the hood,* place the beaker on a steam bath to evaporate the ether. The lecithin may appear as a thick syrup or waxy substance. Remove the beaker occasionally if the mixture starts to splatter. Cool and determine the combined mass of the beaker and lecithin.

B.3 Describe the color and appearance of lecithin. Oxidation may turn it to yellow or brown.

Calculations

B.4 Calculate the mass of the lecithin.

B.5 Determine the percent by mass of lecithin in egg yolk. You already have the mass of the egg yolk in A.6.

$$\text{Mass \% lecithin} = \frac{\text{g lecithin}}{\text{g egg yolk}} \times 100$$

Report Sheet - Lab 31

Date _____   Name _____

Section _____   Team _____

Instructor _____

Pre-Lab Study Questions

1.   How are the structures of triacylglycerols similar to the structures of glycerophospholipids?

     How do they differ?

2.   Draw the structure of the steroid nucleus.

A.   Isolation of Cholesterol from Egg Yolk

A.1   Mass of beaker and egg yolk   _____

        Mass of beaker   _____

A.2   Mass of flask   _____

A.3   Mass of flask and cholesterol   _____

A.4   Color and appearance of cholesterol   _____

A.5   Melting point of crude cholesterol   _____

Calculations

A.6   Mass of egg yolk   _____

A.7   Mass of the cholesterol   _____

A.8   Mass % of cholesterol in egg yolk   _____

        (Show calculations)

Report Sheet - Lab 31

Questions and Problems

Q.1 If a person has high blood cholesterol, why would a nutrionist recommend the use of egg white only in egg dishes?

Q.2 What groups are found on the steroid nucleus in the structure of cholesterol?

B. Isolation of Lecithin from Egg Yolk

B.1 Mass of beaker   _____

B.2 Mass of beaker and lecithin   _____

B.3 Color and appearance of lecithin

Calculations

B.4 Mass of the lecithin   _____

B.5 Mass % of lecithin in egg yolk   _____

(Show calculations)

Questions and Problems

Q.3 Why would lecithin in egg yolk be more useful than a triacylglycerol in the emulsification of oil and water in the preparation of mayonnaise?

Q.4 Lecithins contain a high amount of linoleic acid. Draw the structure of a lecithin containing two linoleic fatty acids

Linoleic acid $CH_3-(CH_2)_4-CH5CH-CH_2-CH5CH-(CH_2)-COOH.$

Saponification and Soaps

Goals

- Prepare soap by the saponification of a fat or oil.
- Observe the reactions of soap with oil, $CaCl_2$, $MgCl_2$, and $FeCl_3$.

Discussion

A. Saponification: Preparation of Soap

For centuries, soaps have been made from animal fats and lye (NaOH), which was obtained by pouring water through wood ashes. The hydrolysis of a fat or oil by a base such as NaOH is called *saponification* and the salts of the fatty acids obtained are called *soaps*. The other product of hydrolysis is glycerol, which is soluble in water.

| Fat (tripalmitin) | Base | Glycerol | Soap (sodium palmitate) |

The fats that are most commonly used to make soap are lard and tallow from animal fat and coconut, palm, and olive oils from vegetables. Castile soap is made from olive oil. Soaps that float have air pockets. Soft soaps are made with KOH instead of NaOH to give potassium salts.

B. Properties of Soaps and Detergents

A soap molecule has a dual nature. The nonpolar carbon chain is hydrophobic and attracted to nonpolar substances such as grease. The polar head of the carboxylate salt is hydrophilic and attracted to water.

The dual polarity of a soap (salt of a fatty acid)

$CH_3CH_2CH_2CH_2CH_2CH_2CH_2CH_2CH_2CH_2CH_2CH_2CH_2CH_2CH_2CH_2 — C — O^-Na^+$

Nonpolar tail (hydrophobic)   Polar head (hydrophilic)

When soap is added to a greasy substance, the hydrophobic tails are embedded in the non-polar fats and oils. However, the polar heads are attracted to the polar water molecules. Clusters of soap particles called *micelles* form with the nonpolar oil droplet in the center surrounded by many polar heads that extend into the water. Eventually all of the greasy substance forms micelles, which can be washed away with water. In hard water, the carboxylate ends of soap react with Ca^{2+}, Fe^{3+}, or Mg^{2+} ions and form an insoluble substance, which we see as a gray line in the bathtub or sink. Tests will be done with the soap you prepare to measure its pH, its ability to form suds in soft and hard water, and its reaction with oils.

Detergents or "syndets" are called synthetic cleaning agents because they are not derived from naturally occurring fats or oils. They are popular because they do not form insoluble salts with ions, which means they work in hard water as well as in soft water. A typical detergent is sodium lauryl sulfate.

$$CH_3(CH_2)_{10}CH_2 - O - \overset{\overset{O}{\|}}{\underset{\underset{O}{\|}}{S}} - O^- Na^+$$

Lauryl sulfate salt,
a nonbiodegradable detergent

As detergents replaced soaps for cleaning, it was found that they were not degraded in sewage treatment plants. Large amounts of foam appeared in streams and lakes that became polluted with detergents. Biodegradable detergents such as an alkylbenzenesulfonate detergent eventually replaced the nonbiodegradable detergents.

$$CH_3(CH_2)_9 - \overset{\overset{CH_3}{|}}{CH} - \bigcirc - \overset{\overset{O}{\|}}{\underset{\underset{O}{\|}}{S}} - O^- Na^+$$

Laurylbenzenesulfonate salt,
a biodegradable detergent

In addition to the sulfonate salts, a box of detergent contains phosphate compounds along with brighteners and perfumes. However, phosphates accelerate the growth of algae in lakes and cause a decrease in the dissolved oxygen in the water. As a result, the lake decays. Some replacements for phosphates have been made.

Lab Information

Time:	2 hr
Comments:	You will be working with hot oil and NaOH. Be sure you wear your goggles.
	Tear out the report sheets and place them beside the matching procedures.
Related Topics:	Esters, saponification, soaps, hydrophobic, hydrophilic

Experimental Procedures

Wear your protective goggles!

A. Saponification: Preparation of Soap

Materials: 150-mL beaker, hot plate, graduated cylinder, stirring rod or stirring hot plate with stirring bar, large watch glass, 400-mL beaker, Büchner filter system, filter paper, plastic gloves, fat (lard, solid shortening, coconut oil, olive or other vegetable oil), ethanol, 20% NaOH, saturated NaCl solution

Weigh a 150-mL beaker. Add about 5 g of fat or oil. Reweigh.

Add 15 mL ethanol (solvent) and 15 mL of 20% NaOH. *Use care when pouring NaOH.* Place the beaker on a hot plate and heat to a gentle boil and stir continuously. A magnetic stirring bar may be used with a magnetic stirrer. Heat for 30 minutes or until saponification is complete and the solution becomes clear with no separation of layers. Be careful of splattering; the mixture contains a strong base. Wear disposable gloves, if available. Do not let the mixture overheat or char. Add 5-mL portions of an ethanol–water (1:1) mixture to maintain volume. If foaming is excessive, *reduce* the heat.

Caution: Oil and ethanol will be hot, and may splatter or catch fire. Keep a watch glass nearby to smother any flames. NaOH is caustic and can cause permanent eye damage. Wear goggles at all times.

Obtain 50 mL of a saturated NaCl solution in a 400-mL beaker. (A saturated NaCl solution is prepared by mixing 30 g of NaCl with 100 mL of water.) Pour the soap solution into this salt solution and stir. This process, known as "salting out," causes the soap to separate out and float on the surface.

Collecting the soap Collect the solid soap using a Büchner funnel and filter paper. See Figure 32.1. Wash the soap with two 10-mL portions of cold water. Pull air through the product to dry it further. Place the soap curds on a watch glass or in a small beaker and dry the soap until the next lab session. Use disposable, plastic gloves to handle the soap. **Handle with care: The soap may still contain NaOH, which can irritate the skin.** Save the soap you prepared for the next part of this experiment. Describe the soap.

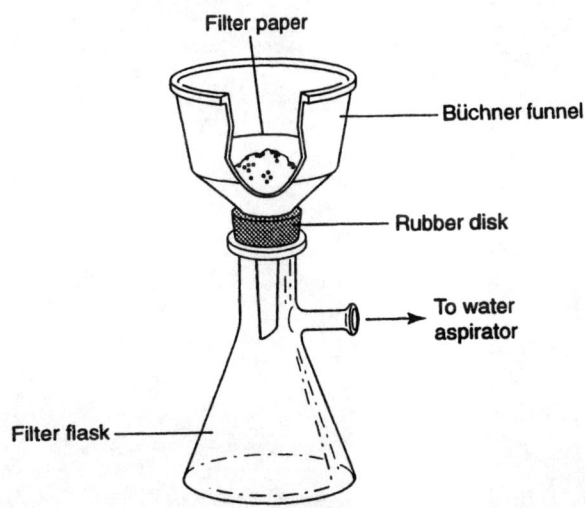

Figure 32.1 Apparatus for suction filtration with Büchner funnel

B. Properties of Soaps and Detergents

Materials: Test tubes, stoppers to fit, droppers, small beakers, 50- or 100-mL graduated cylinder, stirring rod, laboratory-prepared soap (from part A), commercial soap product, detergent, pH paper, oil, 1% $CaCl_2$, 1% $MgCl_2$, and 1% $FeCl_3$

Prepare solutions of the soap you made in part A, a commercial soap, and a detergent by dissolving 1 g of each in 50 mL of distilled water. If the soap is a liquid, use 20 drops.

B.1 **pH test** Place 10 mL of each soap solution in separate test tubes. Use 10 mL of water as a comparison. Label. Dip a stirring rod into each solution, then touch the stirring rod to pH paper. Determine the pH. *Save the tubes for part B.2.*

B.2 **Foam test** Stopper each of the tubes from B.1 and shake for 10 seconds. The soap should form a layer of suds or foam. Record your observations. *Save the tubes for part B.3.*

B.3 **Reaction with oil** Add 5 drops of oil to each test tube from B.2. Stopper and shake each one for 10 seconds. Record your observations. Compare the sudsy layer in each test tube to the sudsy layers in part B.2.

B.4 **Hard water test** Place 5 mL of the soap solutions in three separate test tubes. Add 20 drops of 1% $CaCl_2$ to the first sample, 20 drops of 1% $MgCl_2$ to the second tube, and 20 drops of 1% $FeCl_3$ to the third tube. Stopper each test tube and shake 10 seconds. Compare the foamy layer in each of the test tubes to the sudsy layer obtained in part B.2. Record your observations.

Report Sheet - Lab 32

Date _____   Name _____

Section _____   Team _____

Instructor _____

Pre-Lab Study Questions

1. What happens when a fatty acid is reacted with NaOH?

2. Why is ethanol added to the reaction mixture of fat and base in the making of soap?

3. Why is the product of saponification a salt?

A. Saponification: Preparation of Soap
Describe the appearance of your soap.

Questions and Problems

Q.1 How would soaps made from vegetable oils differ from soaps made from animal fat?

Q.2 How does soap remove an oil spot?

B. Properties of Soaps and Detergents

Tests	Water	Lab Soap	Commercial Soap	Detergent
B.1 pH				
B.2 Foam				
B.3 Oil				
B.4 $CaCl_2$				
$MgCl_2$				
$FeCl_3$				

Report Sheet - Lab 32

Questions and Problems

Q.3 Which of the solutions were basic? Why?

Q.4 Write an equation for the saponification of trimyristin with KOH.

Amines and Amides

Goals

- Draw the structural formulas and give the names of amines.
- Classify amines as primary, secondary, or tertiary.
- Observe some physical properties of amines and amides.
- Write an equation for the formation of an amine salt.
- Write an equation for the formation of an amide and its hydrolysis in acid and base.

Discussion

A. Structure and Classification of Amines

Amines are considered as derivatives of ammonia in which one or more hydrogen atoms are replaced with alkyl or aromatic groups. The number of alkyl groups attached to the nitrogen atom determines the classification of primary, secondary, or tertiary amines.

NH_3	CH_3-N-H (with H above)	CH_3-N-CH_3 (with H above)	CH_3-N-CH_3 (with CH_3 above)
Ammonia	Methylamine (primary, 1°)	Dimethylamine (secondary, 2°)	Trimethylamine (tertiary, 3°)

Amines are often found as part of compounds that are physiologically active or used in medications.

Neo-Synephrine   Histamine   Methamphetamine (methedrine)

B. Solubility of Amines in Water

In water, ammonia and amines with one to four carbon atoms act as weak bases because the unshared pair of electrons on the nitrogen atom attracts protons. The products are an ammonium ion or alkyl ammonium ion and a hydroxide ion.

$$NH_3 \quad + \quad H_2O \quad \longrightarrow \quad NH_4^+ \quad + \quad OH^-$$

Ammonia   Ammonium ion   Hydroxide ion

$$CH_3-NH_2 \quad + \quad H_2O \quad \longrightarrow \quad CH_3-NH_3^+ \quad + \quad OH^-$$

Methylamine   Methylammonium ion   Hydroxide ion

C. Neutralization of Amines with Acids

Because amines are basic, they react with acids to form the amine salt. These amine salts are much more soluble in water than the corresponding amines.

$$CH_3-NH_2 \quad + \quad HCl \quad \longrightarrow \quad CH_3-NH_3^+ \quad + \quad Cl^-$$

Methylamine   Methylammonium chloride
(amine salt)

D. Amides

When a carboxylic acid reacts with ammonia or an amine, the product is an amide. The functional group, called the amide group, and some examples of amides are shown below:

$$\overset{O}{\overset{\|}{-C}}-\overset{|}{N}- \qquad CH_3-\overset{O}{\overset{\|}{C}}-NH_2 \qquad \text{⬡}-\overset{O}{\overset{\|}{C}}-NH_2 \qquad CH_3-\overset{O}{\overset{\|}{C}}-NH-CH_3$$

Amide   Acetamide   Benzamide   *N*-Methylacetamide

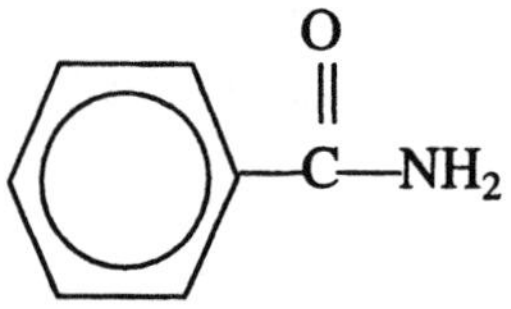

In a reaction called *amidation*, an amide forms when a carboxylic acid is heated with ammonia or an alkyl or aromatic amine.

$$CH_3-\overset{O}{\overset{\|}{C}}-OH \quad + \quad CH_3-NH_2 \quad \longrightarrow \quad CH_3-\overset{O}{\overset{\|}{C}}-NH-CH_3$$

Acetic acid   Methylamine   *N*-Methylacetamide

Hydrolysis of an Amide

When an amide is hydrolyzed, the amide bond is broken and the carboxylic acid and the amine are separated. Hydrolysis takes place in either an acid or a base. Acid hydrolysis produces the carboxylic acid and ammonium salt. In a base, the hydrolysis reaction produces the salt of the carboxylic acid and ammonia. The odor of ammonia and the reaction of ammonia with litmus paper are used to detect the hydrolysis reaction.

$$CH_3-\overset{O}{\overset{\|}{C}}-NH_2 \quad + \quad HCl \quad \longrightarrow \quad CH_3-\overset{O}{\overset{\|}{C}}-OH \quad + \quad NH_4Cl$$

Acetamide   Acetic acid   Ammonium chloride

$$CH_3-\overset{O}{\overset{\|}{C}}-NH_2 \quad + \quad NaOH \quad \longrightarrow \quad CH_3-\overset{O}{\overset{\|}{C}}-O^- \, Na^+ \quad + \quad NH_3(g)$$

Acetamide   Sodium acetate   Ammonia

Lab Information

Time: 2 ½–3 hr

Comments: Some amines have an irritating odor. Work in the hood. Tear out the report sheets and place them next to the matching procedures.

Related topics: Amines, solubility and pH of amines, amidation, amides, hydrolysis of amides

Experimental Procedures

WEAR YOUR PROTECTIVE GOGGLES!

A. Structure and Classification of Amines

Materials: Organic model kits

A.1 Use a model kit to prepare or observe models of ammonia, methylamine, dimethylamine, trimethylamine, and aniline. Write the condensed structural formulas of each model in the laboratory report.

A.2 Classify each amine as primary (1°), secondary (2°), or tertiary (3°).

B. Solubility of Amines in Water

Materials: Aniline, *N*-methylaniline, triethylamine, test tubes, test tube rack, stirring rod, pH paper

WORK IN THE HOOD. THE VAPORS OF AMINES ARE IRRITATING TO THE NOSE AND SINUSES.

B.1 To three separate test tubes, add 5 drops of aniline, *N*-methylaniline, and triethylamine. Draw the condensed structural formulas of each amine and state its classification (1°, 2°, 3°).

B.2 Cautiously note the odor of each. Remember to hold a fresh breath of air while you fan the vapor toward you. Record.

B.3 Add 2 mL of water to each test tube. Stir. Describe their solubility in water.

B.4 Determine the pH of each solution. Dip a stirring rod in each amine solution and then touch it to pH paper. Record. *Save these test tubes and samples for part C.*

C. Neutralization of Amines with Acids

Materials: Test tubes from part B, blue litmus paper, 10% HCl

C.1 Add 10% HCl dropwise to the amine solution until the solution is acidic to litmus paper. Record any changes in solubility of each amine. Note any changes in odor.

C.2 Write and balance equations for the neutralization of aniline, *N*-methylaniline, and triethylamine with HCl.

D. Amides

Materials: Organic model kit, acetamide, benzamide, test tubes, spatula, 10% HCl, 10% NaOH, 250-mL beaker for water bath, hot plate, red litmus paper

D.1 Make a model of acetic acid and ammonia. Show how the structures change in the formation of acetamide. Write the equation for the amidation.

D.2 Place small amounts (tip of a spatula) of acetamide and benzamide in separate test tubes. *Cautiously* note the odor. Add 2 mL of water to each. Record the solubility of each in water.

Acetamide   Benzamide

Hydrolysis of an Amide

D.3 Make a model of acetamide and HCl. Use the models to show the hydrolysis reaction of the amide with HCl, and write the equation.

D.4 Using the test tubes from D.2, add 2 mL of 10% HCl to each. Place the test tubes in a boiling water bath and heat gently for 5 minutes. Cautiously note any odor coming from each mixture. Record your observations.

D.5 Prepare two more test tubes as you did in part D.2. Add 2 mL of 10% NaOH to each. Place the test tubes in a boiling water bath. Wet a piece of pH paper or red litmus paper and hold over the mouth of each test tube. Heat the tubes gently for 5 minutes. Record any change in the color of the litmus paper. Cautiously note any odor coming from each mixture. Record your observations. Write the equation for the hydrolysis of acetamide in base.

Report Sheet - Lab 33

Date _____   Name _____

Section _____   Team _____

Instructor _____

Pre-Lab Study Questions

1. What is the functional group in amines? In amides?

2. What products are formed when amides are hydrolyzed?

3. How is the amide bond important in proteins?

A. Structure and Classification of Amines

Compound	A.1 Condensed Structural Formula	A.2 Classification (1°, 2°, 3°)
Ammonia		
Methylamine		
Dimethylamine		
Trimethylamine		
Aniline		

Questions and Problems

Q.1 In the discussion, the structures are given for Neo-Synephrine and methamphetamine. Give the amine classification of each of the compounds.

Report Sheet - Lab 33

B. Solubility of Amines in Water

	Aniline	*N*-Methylaniline	Triethylamine
B.1 Condensed structural formula			
Classification 1°, 2°, 3°			
B.2 Odor			
B.3 Solubility in water			
B.4 pH			

Questions and Problems

Q.2 What type of compound accounts for the "fishy" odor of fish?

Q.3 Explain why amines are basic.

Report Sheet - Lab 33

C. Neutralization of Amines with Acids

	Aniline	*N*-Methylaniline	Triethylamine
C.1 Solubility after adding HCl			
Odor after adding HCl			

C.2 Equation for the neutralization of aniline with HCl

Equation for the neutralization of *N*-methylaniline with HCl

Equation for the neutralization of triethylamine with HCl

Questions and Problems

Q.4 How does lemon juice remove the odor of fish?

Q.5 Write an equation for the reaction of butylamine with HCl.

Report Sheet - Lab 33

D. Amides

D.1 Equation for the formation of acetamide

	Acetamide	Benzamide
D.2 Odor		
Solubility		

D.3 Equation for the hydrolysis of acetamide in acid.

	Acetamide	Benzamide
D.4 Odor after adding HCl		
D.5 Odor after adding NaOH		
Change in red litmus paper		

Equation for the hydrolysis of acetamide in base.

Questions and Problems

Q.6 You have unknowns that are a carboxylic acide, an ester, and an amine. Describe how you would distinguish among them.

Synthesis of Acetaminophen

Goals

- Write equations for amidation and hydrolysis reactions.
- Prepare the common analgesic acetaminophen.
- Use solubility to purify a crude sample of acetanilide.

Discussion

A. Synthesis of Acetaminophen

Compounds used to relieve pain are called analgesics and compounds used to reduce a fever are antipyretics. Aspirin is both an analgesic and antipyretic and so is acetaminophen, which is an amide.

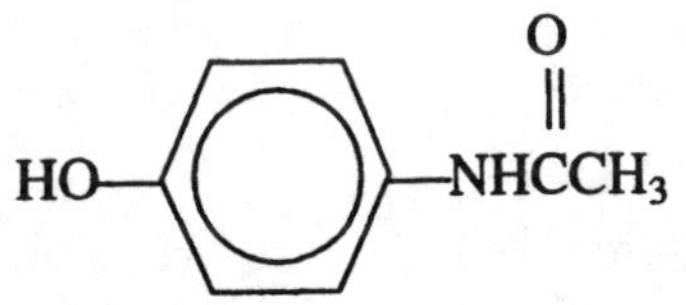

Acetaminophen

People who are sensitive to aspirin may use products such as Tylenol (acetaminophen) or Motrin, Advil, and Nuprin, which contain ibuprofen. Ibuprofen is a carboxylic acid, not an amide.

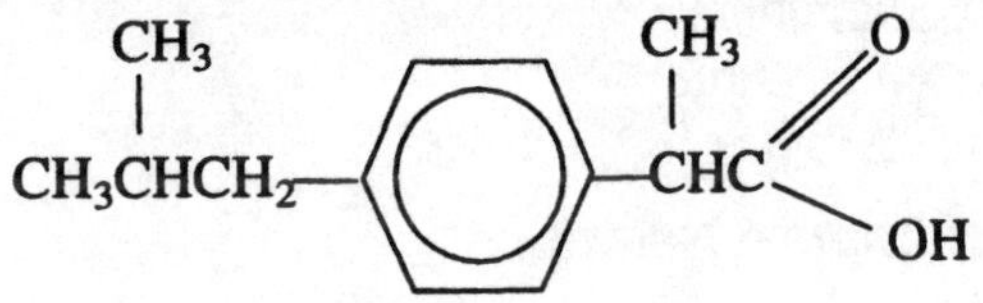

Ibuprofen

Aspirin and acetaminophen are used in several common analgesic preparations, which may also contain caffeine and buffers. See Table 34.1.

Table 34.1 ***Some Products with Aspirin and/or Acetaminophen***

Product	Aspirin	Acetaminophen	Caffeine
Aspirin	325 mg		
Anacin	400 mg		32 mg
Bufferin	324 mg		32 mg
Excedrin	250 mg	250 mg	65 mg
Tylenol		325 mg	
Side effect	Stomach upset; longer bleeding time; possilble toxic levels	Possible liver damage in high dosages for long-term users	Increase in pulse and heart rate

Marketed as Tylenol, acetaminophen is an amide that we can prepare in the laboratory from *p*-aminophenol and a two-carbon group obtained from acetic anhydride.

| *p*-Aminophenol | Acetic anhydride | Acetaminophen | Acetic acid |

Percent Yield

The maximum yield of acetaminophen from 1.5 g of *p*-aminophenol is calculated using the molar mass of the reactant *p*-aminophenol (109 g/mole) and the product acetaminophen (151 g/mole).

$$1.5 \text{ g } p\text{-aminophenol} \times \frac{1 \text{ mole } p\text{-aminophenol}}{109 \text{ g } p\text{-aminophenol}} \times \frac{1 \text{ mole acetaminophen}}{1 \text{ mole } p\text{-aminophenol}} \times \frac{151 \text{ g acetaminophen}}{1 \text{ mole acetaminophen}}$$

$$\text{or } 1.5 \text{ g } p\text{-aminophenol} \times \frac{151 \text{ g acetaminophen}}{109 \text{ g } p\text{-aminophenol}} = \text{g acetaminophen (possible)}$$

The percent yield is calculated by dividing the actual mass of acetaminophen obtained by the maximum possible yield.

$$\% \text{ Yield} = \frac{\text{mass of purified acetaminophen}}{\text{mass of acetaminophen possible}} \times 100\%$$

B. Isolating Acetanilide from an Impure Sample

Acetanilide shows a tenfold increase in solubility from 25°C to 100°C. This difference in solubility can be used to isolate and purify acetanilide from an impure sample.

Acetanilide

The percent yield is calculated by dividing the mass of the purified acetanilide by the mass of the impure sample.

$$\% \text{ Yield} = \frac{\text{mass of purified product}}{\text{mass of impure sample}} \times 100\%$$

Lab Information

Time: 2 ½–3 hr

Comments: When recrystallizing product, use ice water to rinse crystals.
Tear out the report sheets and place them beside the matching procedures.

Related topics: Amides, amidation

Experimental Procedures

 **GOGGLES REQUIRED!**

A. Synthesis of Acetaminophen

Materials: 125-mL Erlenmeyer flask, *p*-aminophenol, 85% H_3PO_4, acetic anhydride, 150-mL beaker, hot plate, stirring rod, ice bath, Büchner filtration apparatus, filter paper, melting point apparatus, watch glass

A.1 Weigh a 125-mL Erlenmeyer flask. Add 1.5 g of *p*-aminophenol and reweigh. *Avoid contact with skin. You may wish to wear gloves.* Add 25 mL of water and 20 drops of 85% H_3PO_4. Then heat the flask on a hot plate to boiling. The *p*-aminophenol should dissolve. Remove the flask from the hot plate. ***Working in the hood, carefully*** add 2 mL of acetic anhydride. Stir. Place the flask in an ice bath. Stir to crystallize the acetaminophen. You may need to scratch the walls of the flask to start the crystallization. If no crystals appear, add a small crystal of acetaminophen to start the formation of solid acetaminophen. Allow the flask to stay in the ice-water bath for 30 minutes.

A.2 Collect the crystals using a Büchner funnel vacuum. (See Lab, "Synthesis of Aspirin," for directions on using the Büchner filtration setup.) Wash the product with 10 mL of ice water. Allow the crystals to dry. Weigh a watch glass. Transfer the crude product to the watch glass and reweigh.

A.3 The melting point (mp) of acetaminophen is 169–171°C; p-aminophenol melts at 189–190°C. If a melting point apparatus is available, determine the melting point of your dry, crude acetaminophen product. Compare your melting point results with the known melting points for the starting and expected products.

Purification of Crude Acetaminophen (Optional)

A.4 Place the crude acetaminophen in a 150-mL beaker. Add 20 mL of water and heat on a hot plate until all of the solid dissolves. If the solution reaches boiling and crystals remain, add more water, a few mL at a time. Remove the flask and allow the solution to cool. When crystals begin to appear, place the flask in an ice bath for 20 minutes. If no crystals appear, scratch the inside walls. Collect the crystals using the Büchner filtration apparatus. Wash with 10 mL of cold water. Transfer the filter paper and crystals to a paper towel and let dry. Weigh a watch glass. Transfer the pure product to the watch glass and reweigh.

A.5 Calculate the percentage yield of the pure acetaminophen.

A.6 Determine the melting point of the recrystallized acetaminophen. How does the mp of the pure product compare with the mp of the crude product?

B. Isolating Acetanilide from an Impure Sample

Materials: Two 250-mL beakers, impure acetanilide, hot plate, short-stem funnel, filter paper, iron ring, Büchner funnel, watch glass, small vial and stopper

B.1   Weigh a 250-mL beaker. Add about 2 g of impure acetanilide. Weigh the beaker and contents. Record. Calculate the mass of the impure sample. Add 50 mL of water to the beaker and heat the mixture on a hot plate until no more solid material appears to dissolve.

While the mixture is heating, heat another beaker of water for use in the filtration. Place a short-stem funnel fitted with filter paper in an iron ring. When you are ready to filter your impure sample, pour some of the hot water through the funnel to warm the glass. Discard the water.

Filter the warm sample of impure acetanilide into a clean beaker or flask. Rinse the funnel with hot water so that crystals do not form in it. Place the beaker or flask containing the warm filtrate in an ice bath. As the filtrate cools, crystals of acetanilide should form.

B.2   Weigh a watch glass. Collect the crystals of acetanilide using the Büchner funnel and suction filtration apparatus. Place the crystals on the weighed watch glass and let them dry. Weigh the watch glass with the dry product. Record the appearance of the purified acetanilide.

B.3   Determine the mass of the acetanilide and calculate the percent yield of the pure product.

B.4   Use a melting point apparatus to determine the melting point of the purified acetanilide. Record.

Optional: Place the purified product in a small vial, stopper, and label with your name, % yield, and melting point of the product. Turn in the product to your instructor.

Report Sheet - Lab 34

Date _____   Name _____

Section _____   Team _____

Instructor _____

Pre-Lab Study Questions

1. Draw the structures of acetaminophen and acetanilide. Describe their similarities.

2. What method is used to remove the impurities in a product?

A. Synthesis of Acetaminophen

A.1 Mass of flask   _____

     Mass of flask and *p*-aminophenol   _____

A.2 Mass of watch glass   _____

     Mass of watch glass and acetaminophen product   _____

A.3. Melting point of crude acetaminophen   _____°C

A.4 Mass of watch glass + pure product   _____

     Mass of watch glass   _____

     Mass of pure product   _____

A.5 Percent yield   _____%
     Show calculations.

A.6 Melting point of purified acetaminophen product   _____°C

     Handbook value for melting point of acetaminophen   _____°C

Questions and Problems

Q.1 Acetaminophen does not act as a base in water, but p-aminophenol does. Explain.

Report Sheet - Lab 34

Questions and Problems

Q.2 What does the comparison of the melting points of your product(s) and the known melting point of acetaminophen tell you about the purity of your product?

Q.3 Phenacetin is another over-the-counter medication for reducing fever and relieving pain.

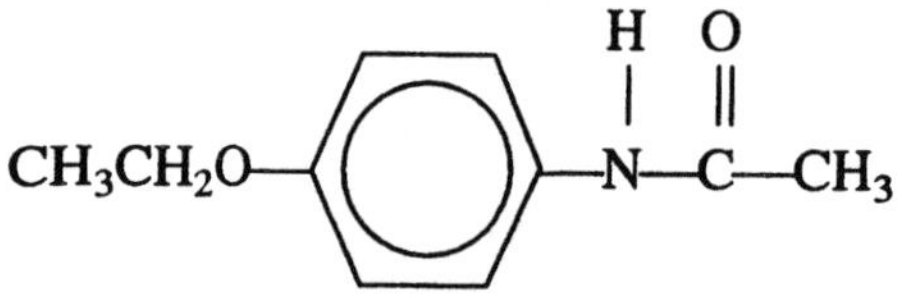

$$CH_3CH_2O-\bigcirc-\overset{H}{\underset{}{N}}-\overset{O}{\underset{}{C}}-CH_3$$

Phenacetin

    a. What are the functional groups in phenacetin?

    b. How might you prepare phenacetin?

B. Isolating Acetanilide from an Impure Sample

B.1 Mass of beaker   _____

    Mass of beaker and impure acetanilide   _____

B.2 Mass of watch glass   _____

    Mass of watch glass and acetanilide product   _____

    Appearance of purified acetanilide   _____

B.3 Mass of purified acetanilide   _____

    Percent yield of the pure product   _____ %
    Show calculations.

B.4 Melting point of acetanilide product   _____ °C

    Handbook value for melting point of acetanilide   _____ °C

Plastics and Polymerization

Goals

- Identify the monomer units in a polymer.
- Write a portion of a polymer from its monomer units.
- Identify the type of polymer in a plastic item from its recycling code.
- Prepare polymers of nylon, polystyrene, and Slime.

Discussion

Polymers are huge molecules that are made by combining many small molecules called monomers. Polymers are prevalent in nature. Cellulose and starch are polymers of glucose, a monosaccharide, and silk and wool as well as the enzymes in our cells, are proteins polymers composed of amino acids. In the last century, scientists have developed many kinds of *synthetic* polymers that are important in our daily lives. Some include the nonstick coating on cooking sheets and pans, foam rubber, disposable diapers, plastic cups, garden hoses, nylon, outdoor clothing and carpeting, plastic wrap, computer disks, and surfboards.

Polyethylene is a common polymer used to make plastic bottles, film, and plastic dinnerware. During polymerization, the double bonds of ethene (ethylene) molecules open and add to the next monomer. When large numbers of monomers combine to form a long carbon chain, it is a polymer of polyethylene.

A. Classification of Plastics

Because many of the plastic synthetic polymers are based on alkanes, the polymers do not decompose easily. They are not biodegradable and contribute to pollution. Recycling programs collect plastic materials and recycle the materials rather than adding them to landfills. Plastic items are labeled on the bottom with a recycling code so they can be sorted according to the type of plastic. (See Table 35.1.)

Table 35.1 *Types of Plastics and Their Recycling Codes*

Recycling Codes	Polymer	Examples
1 PETE	Polyethylene terephthalate	Soda bottles, carpets
2 HDPE	High density polyethylene	Milk and detergent bottles
3 V	Vinyl/Polyvinyl chloride	Plastic pipes, garbage bags
4 LDPE	Low density polyethylene	Soft bottles, carpets, dry cleaners bags
5 PP	Polypropylene	Raincoats, yogurt containers, artificial joints
6 PS	Polystyrene	Coffee cups, foam cartons, toys
7 Other	Resins, mixed polymers	Ketchup bottles

In this lab, we will identify an assortment of plastic items by their recycling codes and look at some of their physical properties. The different types of polymers vary in properties such as density. In recycling plants, plastics are shredded and added to a liquid such as water. Plastics denser than water sink, and those less dense float. Those that sink are recovered separately from those that float.

B. Gluep and Slime®

Slime® is a gel that is popular polymer product enjoyed by children. In the laboratory it is made from polyvinyl alcohol and a saturated borax ($Na_2B_4O_7$) solution. A similar type of gel can be made from Elmer's glue, which contains polyvinyl acetate rather than polyvinyl alcohol. Polyvinyl acetate is a polymer of vinyl acetate

Vinyl alcohol → polymerization → Polyvinyl alcohol

Vinyl acetate → polymerization → Polyvinyl acetate

When borate from Borax is added to polyvinyl alcohol or polyvinyl acetate, large numbers of cross links form a viscous substance. The number of cross links, which are hydrogen bonds, can be increased or decreased by changing the borate concentration. They also break and reform with handling or from the weight of the gel. Thus the gel tends to flow, change shape, and break if it is pulled apart suddenly. Adding a few drops of acid destroys the gel; adding some NaOH reforms the gel.

C. Polystyrene

Styrene is the compound that is polymerized to give polystyrene

Styrene → Polystyrene section

Polystyrene is a clear, brittle plastic used to make plastic glasses and coffee cups. In the polystyrene polymer, there may be as many as 3000 styrene monomers. At high temperatures, the polymerization of styrene is spontaneous, but at lower temperature an initiator such as benzoyl peroxide, which forms a high-energy radical, is required. You may be familiar with the compound benzoyl peroxide, which is used in many skincare creams for treatment of acne.

To prevent spontaneous polymerization of styrene in the bottle on a shelf, the manufacturer adds a small amount of an inhibitor, which is 4-*tert*-butylcatechol. This inhibitor is removed by passing styrene through some alumina, which absorbs the inhibitor.

Benzoyl peroxide   Benzoyloxy radial initiator

D. Nylon

Nylon was introduced in 1938, which makes it one of the first synthetic polymers. Polymers known as nylon are polyamides made by the condensation of diamines and dicarboxylic acids. Nylon 6,6 is the condensation of hexamethylene diamine and adipic acid and nylon 6,10 uses hexamethylene diamine and sebacic acid. Two numbers indicating the number of carbon atoms designate the different nylons, the first for the diamine and the second for the dicarboxylic acid. Thus in nylon 6,10 the diamine has 6 carbons, and the dicarboxylic acid has 10 carbons

In this experiment, sebacoyl chloride rather than sebacic acid is heated with the amine. The two reactants will be present in two layers. The polymer will form a film at the interface where the diamine and the diacid chloride are in contact.

Lab Information

Time: 2-3 hr
Comments: Tear out the Lab report sheets and place them beside the matching procedures.
Related topics: monomers, polymers, and polymerization

Experimental Procedures

WEAR YOUR PROTECTIVE GOGGLES!

A.   Classification of Plastics

Materials: samples of plastic items such as Styrofoam cups, Saran wrap, milk cartons, nylon, transparencies, garden hose, plastic glasses (smash with a hammer), yogurt containers, bubble wrap, an alcohol-water mixture (9 mL 70% isopropyl alcohol (rubbing alcohol) and 6 mL water), stirring rods, forceps, acetone, two 250-mL beakers, test tubes, and test tube rack, small pieces of each type of plastic, unknowns (optional), Bunsen burner

A.1 List the type of item and its use. Record its recycling code. Write the name of the type of plastic. Using a reference such as your textbook, write the name of the monomer for the plastic.

A.2 *Density Test*
Fill a 250-mL beaker about $^1/_2$ full with water. Fill a second 250-mL beaker about $^1/_2$ full with an isopropyl alcohol:water (3:2) mixture. Obtain 2 small pieces of each type of plastic. Add one piece to each beaker. If the plastic piece floats, use a stirring rod to gently push the plastic piece below the surface. Record whether each type of plastic is more dense or less dense than water (1.0 g/mL) and more dense or less dense than water-alcohol mixture.

Obtain an unknown piece of plastic. Describe its properties. Determine its results in the density test. Identify the types of plastic it may be.

A.3 *Solubility In Acetone Test*   **This may be a demonstration set up in the hood.**
Caution: Acetone is flammable. Work in the hood. Obtain a sample of each type of plastic. Place 4 mL of acetone in each of 6 test tubes in a test tube rack. Add a piece of one type of plastic to each test tube. Allow the plastic piece to remain in the acetone for 10 minutes. Use forceps to remove the plastic piece and rinse with water. Press between your fingers. Some plastics "swell" in organic solvents and become soft and pliable. Compare the plastic pieces before and after acetone treatment. Record any changes in the appearance of each type of plastic.

Test another piece of the same unknown plastic as in A.2. Identify the types of plastic it may be. Discard in a waste container. Discard the acetone in a waste acetone container.

A.4 *Combustion Test*   **Work in the hood**
Obtain small pieces of each type of plastic. Using forceps, hold a piece in the flame of a Bunsen burner to determine if the plastic ignites. If it begins to burn, remove it from the flame and allow it to continue to burn. Observe how fast it burns and the color of its flame and residue.
Test another piece of your same unknown plastic. Identify the types of plastic it may be. Place the plastic residue in a waste container.

B.   Gluep and Slime®

Materials: 10-mL and 50-mL graduated cylinders, Styrofoam cup, plastic sticks or spatulas, saturated borate solution, Elmer's glue, lab gloves, 4 % polyvinyl alcohol solution

Gluep

B.1   Obtain a Styrofoam cup. Place 20 mL of Elmer's glue and 20 ml of water in the cup. Sir to mix. Add 10 of the borate solution to the cup while stirring vigorously. Continue to stir for at least 10 minutes until it is a viscous gel. Wearing gloves, remove the gel and knead it like bread for 3-4 minutes.

Observe the physical properties of the gel. What does it smell like? What is its texture? What happens when you stretch it slowly? Fast? What happens if you roll it into a ball and let it sit on a piece of paper on the lab bench? Record your observations. What happens if you make a ball of the gel and drop it onto the lab bench? What happens if you roll the gel into a long, thin roll and pull the ends apart?

The following may be done by three lab teams each choosing one procedure.

B.2   Repeat procedure B.1 but use 40 mL of water.

B.3   Repeat procedure B.1 but do not add water.

B.4   Repeat procedure B.1 but use 20 mL of borate solution.

Slime®

B.5   Obtain a Styrofoam cup. Add 20 mL of 4% polyvinyl alcohol solution and 5 mL of the borate solution. Stir for 5 minutes. Remove and knead like bread. Compare the properties of Slime to the Gluep product.

C.   Polystyrene

Materials: Styrene, benzoyl peroxide (or an acne preparation which contains 5% or10% benzoyl peroxide), stirring rods, alumina, funnel, filter paper, hot plate, heavy-duty aluminum foil, 10- or 20-mL beaker, 50-mL beaker, wood stick

1. Place 0.3-0.4 g of alumina in a test tube. Carefully add 4 mL of styrene, which contains inhibitor.

2. Fold a filter paper. Pour the styrene-alumina mixture into the filter paper and collect the styrene (inhibitor removed) in a small beaker. Place used alumina in a nonhazardous waste container.

3. Obtain two square sheets of aluminum foil 10 cm × 10 cm. Using a double layer of the aluminum sheets, form a mold by molding them around the bottom of a small beaker. Place this mold in a 50-mL beaker.

4. Prepare a boiling water bath using a hot plate.

5. Weigh out 0.050 g of benzoyl peroxide (or 0.2 g of a 10% benzoyl peroxide acne cream preparation) and add to the styrene (filtered). Swirl the beaker 3-4 minutes. Pour the mixture into the aluminum foil mold in the 50-mL beaker. Clamp the beaker containing the foil mold and styrene mixture to the ring stand. Lower the bottom of the beaker into the boiling water bath. Be careful not to spill the styrene mixture out of the foil mold as you set this up.

Leave the beaker in the water bath for about 45 minutes. Add more water to the water bath as needed. Occasionally stir the styrene with a wooden stick. As the temperature rises, the mixture should become more viscous. *During this heating time, proceed to the next lab, which is the synthesis of nylon.*

6. Remove the beaker and the aluminum mold with the styrene product from the water bath, and allow it to cool slowly to room temperature. Then place the beaker in an ice bath for 20 minutes. If you leave the wooden stick in the styrene, the styrene will solidify around it. Remove the foil mold from the beaker, and separate the aluminum foil from around the polystyrene. You may need to let the polymer harden until the next laboratory time.

C.1 Describe the appearance and texture of the polystyrene product.

D. Nylon

This experiment may be done in pairs of lab teams.

Caution: Use a fume hood. The reactants are irritating to the skin and eyes.

Hexamethylenediamine and sebacoyl chloride are irritating to the skin, eyes, and respiratory system. Sodium hydroxide is extremely caustic and can cause severe burns. Contact with the skin and eyes must be prevented. Hexane is extremely flammable. Hexane vapor can irritate the respiratory tract and, in high concentrations, be narcotic.

Materials: 50-mL, 100-mL, and 250 beakers, 10-mL and 50-mL graduated cylinders, forceps, metal spatula, large test tube (for spooling the nylon), stirring rods, gloves (must not dissolve in hexane), 50% aqueous ethanol solution, 6 M HCl, 6 M NaOH

Solution 1: 4% hexamethylenediamine and NaOH. If solid, place the reagent bottle in hot water to melt (mp 39°C).

Solution 2: 4% sebacoyl chloride, $ClCO(CH_2)_8COCl$, in hexane

1. Obtain a clean 50-mL beaker. Add 20 mL of solution 1 (1,6-hexamethylenediamine and NaOH). Using a 50-mL graduated cylinder, obtain 20 mL of solution 2 (4% sebacoyl chloride in hexane).

> *Put on gloves. Work in the hood*

2. Tilt the beaker containing the hexamethylenediamine solution and slowly pour the sebacoyl chloride solution down the wall of the beaker. Allow the mixture to site undisturbed for 1 minute. Two layers will form with a whitish film of nylon polymer film at the interface. Run a metal spatula around the walls of the beaker to loosen the film from the beaker sides.

3. Carefully lower the tip of a pair of forceps through the top layer to take hold of the center of the nylon film. Slowly pull a strand of the polymer upward out of the solution. A continuous section of nylon should form. Secure the end of the strand in the forceps by wrapping it around the center of the large test tube. Holding the test tube, turn it slowly to draw out a continuous nylon strand from the interface. You should be able to draw out most of the interface film. If the strand breaks, use forceps to start a new one. Rinse the nylon strand on the test tube under tap water to remove excess reactants. *Do not touch the nylon strand before you have thoroughly washed it with water.*

4. Use a metal spatula to detach the nylon loop from the test tube and slide the nylon loop into a beaker containing 20 mL of 50% aqueous alcohol solution to remove any remaining reactants. Rinse with water again. Using forceps grab an end of the nylon and stretch out the nylon strand and place it on paper towels until it is dry.

D.1 Describe the appearance and texture of the nylon polymer.

D.2 Cut a 15-20 cm section of nylon and pull on both ends. When you stretch the nylon, does it break or return to its original form (elastic)? Record your observations.

D.3 Cut several 2-3 cm sections of nylon. Place one piece in a test tube containing 5 mL of 6M HCl. Place a second piece in a test tube containing 5 mL of 6M NaOH. Place a third piece in a test tube containing 5 mL of acetone. Observe any changes in the nylon in the presence of a strong acid, a strong base, and an organic solvent. Record your observations.

D.4 Light a Bunsen burner and use forceps to hold the nylon piece in the flame. Record your observations on the flammability of nylon.

Disposal

Stir the remaining reaction mixture to form a ball of nylon and discard in the proper solid waste container. Use a metal spatula to remove any remaining nylon polymer from the glass. Place remaining solvents in the proper waste liquid container. Dispose of acid and base solution by pouring them into the sink and rinsing with plenty of water. Dispose of acetone in the proper waste container.

Report Sheet - Lab 35

Date _____   Name _____

Section _____   Team _____

Instructor _____

Pre-Lab Study Questions

1. Why do most plastics remain in landfills for a long time?

2. What is the purpose of the recycling codes on plastics?

A. Classification of Plastics

A.1 Plastic Item/Use	Recycling code	Name of plastic	Monomer unit (name)

Type of Plastic	A.2 Water	A.2 Water-Alcohol	A.3 Acetone	A.4 Combustion
PETE				
HDPE				
V				
LDPE				
PP				
PS				
Unknown				
Possible plasic types				

What type(s) of plastic is your unknown? _____

Report Sheet - Lab 35

Questions and Problems

Q.1 If you are going to make life preservers, which types of plastic could you use? Why?

Q.2 Which type(s) of plastic would you not use for storing left over acetone? Why?

Q.3 You are going into a small business to manufacture handles for a barbecue. Which types of plastic could you use?

B. Gluep and Slime®

Properties	Gluep Samples				Slime®
	B.1	B.2	B.3	B.4	B.5
Smell and texture					
Stretching slowly and then fast					
Letting it sit on the lab bench					
Dropping it on the lab bench					
Pulling it apart					
Other					

Report Sheet - Lab 35

Questions and Problems

Q.4 What combination of reactants for Gluep was the best to work with? Why?

Q.5 What combination of reactants for Gluep did not work well?

Q.6 How does the "slime" product compare to your Gluep product?

C. Polystyrene

C.1 Appearance and texture of your polystyrene sample.

D. Nylon

D.1 Appearance and texture of the nylon polymer.

D.2 Behavior of nylon when stretched.

D.3 Behavior of nylon in acid, base, and organic solvent

HCl	NaOH	Acetone

D.4 Describe the flammability of nylon.

Goals

- Use R groups to determine if an amino acid will be acidic, basic, or neutral; hydrophobic or hydrophilic.
- Use paper chromatography to separate and identify amino acids.
- Calculate R_f values for amino acids.

Discussion

A. Amino Acids

In our body, amino acids are used to build tissues, enzymes, skin, and hair. About half of the naturally occurring amino acids, the *essential amino acids,* must be obtained from the proteins in the diet because the body cannot synthesize them. Amino acids are similar in structure because each has an amino group ($-NH_2$) and a carboxylic acid group ($-COOH$). Individual amino acids have different organic groups *(R groups)* attached to the alpha carbon atom. Variations in the R groups determine whether an amino acid is hydrophilic or hydrophobic, and acidic, basic, or neutral.

Some R groups contain carbon and hydrogen atoms only, which makes the amino acids nonpolar and hydrophobic ("water-fearing"). Other R groups contain OH or SH atoms and provide a polar area that makes the amino acids soluble in water; they are hydrophilic ("water-loving"). Other hydrophilic amino acids contain R groups that are carboxylic acids (acidic) or amino groups (basic). The R groups of some amino acids used in this experiment are given in Table 36.1.

Table 36.1   *Amino Acids Found in Nature*

R Group	Amino Acid	Symbol	Polarity	Reaction to Water
H—	Glycine	Gly	Nonpolar	Hydrophobic
CH_3—	Alanine	Ala	Nonpolar	Hydrophobic
(⬡)—CH_2—	Phenylalanine	Phe	Nonpolar	Hydrophobic
HO—CH_2—	Serine	Ser	Polar	Hydrophilic
$\overset{O}{\overset{\|\|}{HOC}}$—$CH_2$—	Aspartic acid	Asp	Polar, acidic	Hydrophilic
$\overset{O}{\overset{\|\|}{HOC}}$—$CH_2$—$CH_2$—	Glutamic acid	Glu	Polar, acidic	Hydrophilic
$H_2NCH_2CH_2CH_2CH_2$—Lysine		Lys	Polar, basic	Hydrophilic

Ionization of Amino Acids

An amino acid can ionize when the carboxyl group donates a proton, and when the lone pair of electrons on the amino group attracts a proton. Then the carboxyl group has a negative charge, and the amino group has a positive charge. The ionized form of an amino acid, called a *zwitterion* or *dipolar ion,* has a net charge of zero.

$$CH_3(CH_2)_{10}CH_2 - O - \overset{\overset{\displaystyle O}{\parallel}}{\underset{\underset{\displaystyle O}{\parallel}}{S}} - O^- Na^+$$

In acidic solutions (low pH), the zwitterion *accepts* a proton (H^+) to form an ion with a positive charge. When placed in a basic solution (high pH), the zwitterion *donates* a proton (H^+) to form an ion with a negative charge. This is illustrated using alanine.

$$\underset{\substack{\text{High pH} \\ \text{(charge = 1−)}}}{NH_2 - \overset{\overset{\displaystyle CH_3}{|}}{CH} - \overset{\overset{\displaystyle O}{\parallel}}{C} - O^-} \quad \xleftarrow{\text{Base}} \quad \underset{\substack{\text{Zwitterion (neutral pH)} \\ \text{(charge = 0)}}}{\overset{+}{NH_3} - \overset{\overset{\displaystyle CH_3}{|}}{CH} - \overset{\overset{\displaystyle O}{\parallel}}{C} - O^-} \quad \xrightarrow{\text{Acid}} \quad \underset{\substack{\text{Low pH} \\ \text{(charge = 1+)}}}{\overset{+}{NH_3} - \overset{\overset{\displaystyle CH_3}{|}}{CH} - \overset{\overset{\displaystyle O}{\parallel}}{C} - OH}$$

Donates H+   accepts H+

B. Chromatography of Amino Acids

Chromatography is used to separate and identify the amino acids in a mixture. Small amounts of amino acids and unknowns are placed along one edge of Whatman #1 paper. The paper is then placed in a container with solvent. With the paper acting like a wick, the solvent flows up the chromatogram, carrying amino acids with it. Amino acids that are more soluble in the solvent will move higher on the paper. Those amino acids that are more attracted to the paper will remain closer to the origin line. After removing and drying the paper, the amino acids can be detected (visualized) by spraying the dried chromatogram with ninhydrin.

The distance each amino acid travels up the paper from the origin (starting line) is measured and the R_f values calculated. The R_f value is the distance traveled by an amino acid compared to the distance traveled by the solvent. See Figure 36.1.

$$R_f = \frac{\text{distance traveled by an amino acid (cm)}}{\text{distance traveled by the solvent (cm)}}$$

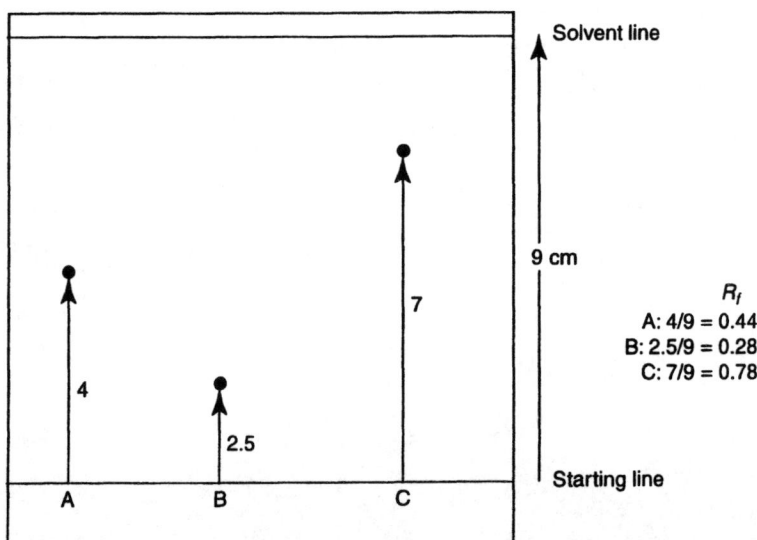

Figure 36.1 A developed chromatogram (R_f values calculated for A, B, and C)

To identify an unknown amino acid, its R_f value and color with ninhydrin is compared to the R_f values and colors of known amino acids in that solvent system. In this way, the amino acids present in an unknown mixture of amino acids can be separated and identified.

Lab Information

Time: 2 ½ –3 hr
Comments: Ninhydrin spray causes stains. Use it carefully.
                       Tear out the report sheets and place them next to the matching procedures.
Related Topics: Amino acids, zwitterions

Experimental Procedures

GOGGLES REQUIRED!

A. Amino Acids

Materials: Organic model kits or prepared models

A.1 Using an organic model kit, construct models of glycine and alanine. Draw their structures. Convert the alanine model to a model of serine. Indicate whether each of the amino acids would be hydrophobic or hydrophilic.

A.2 Form ionized (zwitterion) glycine by removing a H atom from the –COOH group and placing it on the N atom in the –NH_2 group. Draw the structure of the glycine zwitterion.

A.3 Write the structure of glycine in a base and in an acid.

B. Chromatography of Amino Acids

Materials: 600-mL beaker, plastic wrap, plastic gloves, Whatman chromatography paper
                       #1 (12 cm × 24 cm), toothpicks or capillary tubing, drying oven (80°C) or hair
                       dryer, metric ruler, stapler, amino acids (1% solutions): phenylalanine, alanine, glu
                       tamic acid, serine, lysine, aspartic acid, and unknown
                       Chromatography solvent: isopropyl alcohol, 0.5 M NH_4OH; 0.2% ninhydrin spray

Preparation of paper chromatogram Using forceps or plastic gloves (or a sandwich bag), pick up a piece of Whatman #1 chromatography paper that has been cut to a size of 12 cm × 24 cm. *Keep your fingers off the paper because amino acids can be transferred from the skin.* When this paper is rolled into a cylinder, it should fit into the chromatography tank (large beaker) without touching the sides. Draw a pencil (lead) line about 2 cm from the long edge of the paper. This will be the starting or origin line. Mark off seven points about 3 cm apart along the line. (See Figure 36.2.) Place your name or initials in the upper corner with the pencil.

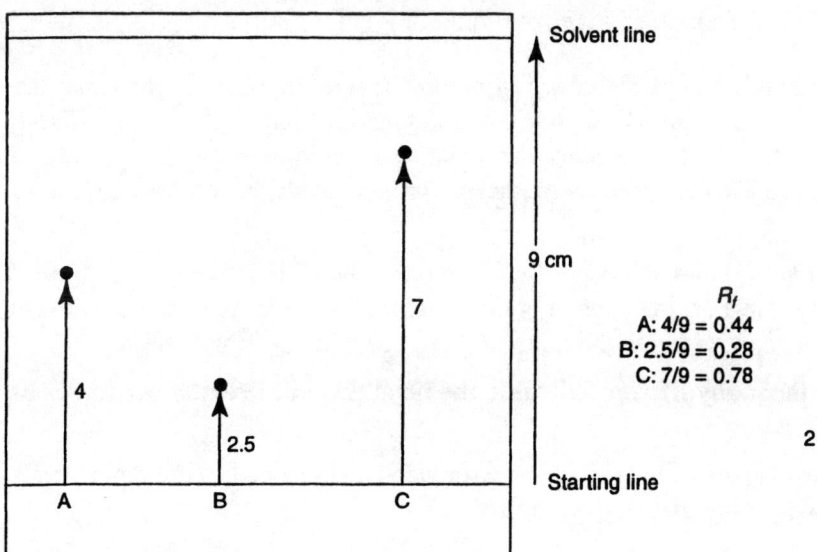

Figure 36.2 Preparation of a chromatogram

Application of amino acids Apply small amounts of the following 1% amino acid solutions: phenylalanine, alanine, glutamic acid, serine, lysine, and aspartic acid. Also prepare a spot of an unknown. Use the toothpick applicators or capillary tubes provided in each amino acid solution to make a small spot (the size of the letter o) by lightly touching the tip to the paper. After the spot dries, retouch the spot one or two more times to apply more amino acid, but keep the diameter of the spot as small as possible. A hair dryer can be used to dry the spots. *Always return the applicator to the same amino acid solution.* Using a pencil, label each spot as you go along. Allow the spots to dry.

Preparation of chromatography tank *Work in the hood.* Prepare the solvent by mixing 10 mL of 0.5 M NH$_4$OH and 20 mL of isopropyl alcohol. Pour the solvent into a 600-mL beaker to a depth of about 1 cm but not over 1.5 cm. (The height of the solvent must not exceed the height of the origin line on your chromatography paper.) Cover the beaker tightly with plastic wrap. This is your chromatography tank. Label the beaker with your name and leave it in the hood.

Running the chromatogram Roll the paper into a cylinder and staple the edges *without overlapping. The edges should not touch.* Slowly lower the cylinder into the solvent of the chromatography tank with the row of amino acids at the bottom. Make sure that the paper does not touch the sides of the beaker. See Figure 36.3. Cover the beaker with the plastic wrap and leave it undisturbed. The tank must not be disturbed while solvent flows up the paper. Let the solvent rise until it is 2–3 cm from the top edge of the paper. It may take 45–60 minutes. *Do not let the solvent run over the top of the paper.*

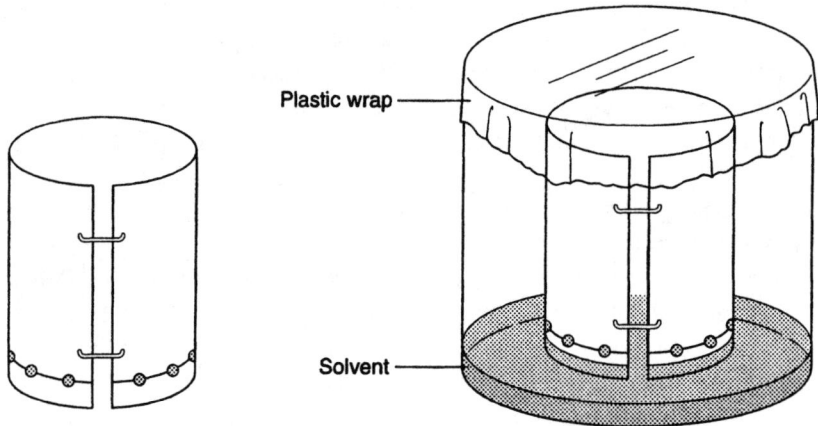

Plastic wrap

Solvent

Figure 36.3 Chromatogram in a solvent tank

Visualization of amino acids *Working in the hood*, carefully remove the paper from the tank. Take out the staples and spread the chromatogram out on a paper towel. *Immediately* mark the solvent line with a pencil. Allow the chromatogram to dry completely. A hair dryer or an oven of about 80°C may be used to speed up the drying process. Pour the solvent into a waste solvent container.

Working in the hood, spray the paper lightly, but evenly, with a ninhydrin solution. Dry the sprayed paper by placing it in a drying oven at about 80°C for 3–5 minutes or use a hair dryer. Distinct, colored spots will appear where the ninhydrin reacted with the amino acids.

Caution: Use the ninhydrin spray inside the hood. Do not breathe the fumes or get spray on your skin.

B.1 Draw the chromatogram on the report sheet, or staple the original to the report sheet. Record the color of each spot on the drawing or original.

B.2 Measure the distance (cm) from the starting line to the top of the solvent line to obtain the distance traveled by the solvent.

B.3 Outline each spot with a pencil. Place a dot at the center of each spot. Measure the distance in centimeters (cm) from the origin to the center dot of each spot.

B.4 Calculate and record the R_f values for the known amino acid samples and the unknown amino acid.

$$R_f = \frac{\text{distance traveled by an amino acid}}{\text{distance traveled by the solvent}}$$

B.5 **Identification of unknown amino acids** Compare the color and R_f values produced by the unknown amino acids. Identical amino acids will give similar R_f values and form the same color with ninhydrin. Identify the amino acid(s) in the unknown.

Report Sheet - Lab 36

Date _____   Name _____

Section _____   Team _____

Instructor _____

Pre-Lab Study Questions

1. What are the functional groups in all amino acids?

2. How does an R group determine if an amino acid is acidic, basic, or nonpolar?

A.   Amino Acids

A.1	Structures of Amino Acids	
Glycine	**Alanine**	**Serine**
Hydrophobic or hydrophilic?		

A.2 Zwitterion structure of glycine	A.3 Glycine ion in base	Glycine ion in acid

Report Sheet - Lab 36

Questions and Problems

Q.1 Write the structure of the zwitterion of alanine.

Q.2 Write the prevalent form of alanine in an acidic solution.

B. Chromatography of Amino Acids

B.1 Chromatogram drawing or original, with colors of spots written in

Calculation of R_f Values

B.2 Distance from origin to solvent line: _____

Amino acid	Color	B.3 Distance (cm) amino acid traveled	B.4 R_f value
Phenylalanine			
Alanine			
Glutamic acid			
Serine			
Lysine			
Aspartic acid			
Unknown			

B.5 Identification of unknown # _____ : _____

Experimental Procedures

A. Peptide Bonds

 **GOGGLES REQUIRED!**

Materials: Organic model set

A.1 Make models of glycine and serine. Remove the components of water (H—OH) from an amino and carboxylic acid group to form the dipeptides glycylserine and serylglycine. Draw each of their structures.

A.2 Demonstrate hydrolysis of the dipeptide serylglycine by breaking the peptide bond and adding the components of H_2O. Write a balanced equation for the hydrolysis.

B. Structure of Proteins

Materials: Organic model set

B.1 Make a model of an amino acid. Draw its structure and name it. Share your model with two other students, or groups of students who will make models of other amino acids. Line the models up and write their amino acid structures.

B.2 Form peptide bonds between the amino acids by removing the components of water. Write an equation for the formation of a tripeptide. Use the symbols for the amino acids to write the tripeptide order. Rearrange the amino acids and prepare a different tripeptide. Write the structure and name of the new tripeptide.

C. Denaturation of Proteins

Materials: Test tubes, test tube holder, 10-mL graduated cylinder, 1% egg albumin (or an egg, cheesecloth, and beaker), 10% HNO_3, 10% NaOH, 95% ethyl alcohol, 1% $AgNO_3$ (dropper bottle)

A fresh egg albumin solution can be prepared by mixing the white from one egg with 200 mL of water and filtering the mixture through cheesecloth into a beaker.

Place 2–3 mL of egg albumin solution in each of five test tubes. Use one sample for each of the following tests. Record your observations and give a brief explanation for the results.

C.1 **Heat** Using a test tube holder, heat the egg albumin solution over a low flame. Describe any changes in the solution.

C.2 **Acid** Add 2 mL of 10% HNO_3.

C.3 **Base** Add 2 mL of 10% NaOH.

C.4 **Alcohol** Add 4 mL of 95% ethyl alcohol. Mix.

C.5 **Heavy metal ions** Add 10 drops of 1% $AgNO_3$.

D. Isolation of Casein (Milk Protein)

Materials: 150-mL beaker, hot plate or Bunsen burner, thermometer, funnel or Büchner
filtration apparatus, filter paper, watch glass, nonfat milk, 10% acetic acid, dropper,
pH paper, stirring rod

D.1 Weigh a 150-mL beaker. Add about 20 mL of nonfat milk to the beaker and weigh. Calculate the
mass of the nonfat milk sample.

D.2 Using pH indicator paper, determine the pH of the milk sample.

D.3 Warm the sample on a hot plate or a Bunsen burner until the temperature reaches about 50°C.
Remove the beaker and milk from the heat and add 10% acetic acid, drop by drop. You may need
2–3 mL. Stir continuously. At the isoelectric point, the casein (milk protein) becomes insoluble.
When no further precipitation occurs, stop adding acid. If the liquid layer is not clear, heat the
mixture gently for a few more minutes. Determine the pH at which the casein becomes insoluble in
solution. This is the pH of the isoelectric point of casein.

D.4 Collect the solid protein using a funnel and filter paper or the Büchner filtration apparatus. Wash
the protein with two 10-mL portions of water. Weigh a watch glass. Transfer the protein to the
watch glass and let the protein dry. Weigh. Calculate the mass of milk protein. *Save for part E.*

D.5 Calculate the percentage of casein in the nonfat milk.

$$\% \text{ Casein } = \frac{\text{mass (g) of casein}}{\text{mass (g) of milk}} \times 100\%$$

E. Color Tests for Proteins

Materials: Test tubes, test tube rack, 10-mL graduated cylinder, boiling water bath, cold water
bath, pH paper, spatula, dropper bottles of 1% amino acid solution (glycine,
tyrosine), dropper bottles of 1% solutions of proteins (gelatin, egg albumin), casein
from part D, 0.2% ninhydrin solution, concentrated HNO_3 (dropper bottle), 10%
NaOH, 5% $CuSO_4$ (biuret), red litmus paper

E.1 **Biuret test** In four separate test tubes, place 2 mL of solutions of glycine, tyrosine, gelatin, and
egg albumin. To the fifth tube, add a small amount of the solid casein (from part D), the amount
held on the tip of a spatula. To each sample, add 2 mL of 10% NaOH and stir. Then add 5 drops of
biuret reagent (5% $CuSO_4$), and stir. Record the color of each sample. The formation of a
pink-violet color indicates the presence of a protein with two or more peptide bonds. If such a
protein is not present, the blue color of the cupric sulfate will remain (negative result). Record
the results and your conclusions.

E.2 **Ninhydrin test** In four separate test tubes, place 2 mL of the solutions of glycine, tyrosine,
gelatin, and egg albumin. To the fifth tube, add a small amount of the solid casein (from part D),
the amount held on the tip of a spatula. Add 1 mL of 0.2% ninhydrin solution to each sample.
Place the test tubes in a boiling water bath for 4–5 minutes. Look for the formation of a blue-violet
color. Record your observations.

E.3 **Xanthoproteic test** *(optional)* Place 1 mL of the solutions of glycine, tyrosine, gelatin, and egg
albumin in four test tubes. To a fifth tube, add a small amount of casein (from part D). *Cautiously*
add 10 drops of concentrated HNO_3 to each sample. Place the test tubes in a boiling water bath
and heat for 3–4 minutes. Remove the test tubes, place them in cold water, and let them cool.
Carefully add 10% NaOH, drop by drop, until the solution is just basic (turns red litmus blue).
This may required 2–3 mL of NaOH. *Caution: heat will be evolved.* Look for the formation of a
yellow-orange color, which may vary in intensity. Record your observations.

Report Sheet - Lab 37

Date _____   Name _____

Section _____   Team _____

Instructor _____

Pre-Lab Study Questions

1. What is a peptide bond?

2. How does the primary structure of proteins differ from the secondary structure?

A. Peptide Bonds

A.1   Structure of glycylserine

Structure of serylglycine

A.2   Hydrolysis of serylglycine

Report Sheet - Lab 37

B. Structure of Proteins

B.1 Amino acid structures and names

B.2 Equation for the formation of the first tripeptide

Order of amino acids using symbols

Equation for the formation of the second tripeptide

Order of amino acids using symbols

Report Sheet - Lab 37

C. Denaturation of Proteins

Treatment	Observations of Egg Albumin	Explanation
C.1 Heat		
C.2 Acid		
C.3 Base		
C.4 Alcohol		
C.5 Heavy metal ions		

Questions and Problems

Q.1 Why are heat and alcohol used to disinfect medical equipment?

Q.2 Why is milk given to someone who accidentally ingests a heavy metal ion such as silver or mercury?

Report Sheet - Lab 37

D. Isolation of Casein (Milk Protein)

D.1 Mass of beaker	
Mass of beaker and milk	
Mass of milk	
D.2 pH of milk	
D.3 pH when casein precipitates	
D.4 Mass of watch glass	
Mass of watch glass and casein	
Mass of casein	
D.5 Percent casein *Show calculations.*	

Questions and Problems

Q.3 Compare the pH of the milk sample and the pH at which the casein solid forms.

Q.4 How does a change in pH affect the structural levels of a protein?

Report Sheet - Lab 37

E. Color Tests for Proteins

Observations of Color Tests			
Sample	E.1 Biuret	E.2 Ninhydrin	E.3 Xanthoproteic
Glycine			
Tyrosine			
Gelatin			
Egg albumin			
Casein (milk protein)			

Questions and Problems

Q.5 After working with HNO_3, a student noticed that she had a yellow spot on her hand. What might be the reason?

Q.6 Which samples give a negative biuret test? Why?

Q.7 What functional group gives a positive test in the xanthoproteic test?

Q.8 What tests could you use to determine whether an unlabeled test tube contained an amino acid or a protein?

Goals

- Prepare a solution of the enzyme amylase.

- Describe the role of an enzyme as a catalyst in biological systems.

- Set up chemical tests that indicate the rate of an enzyme-catalyzed reaction.

- Observe the effects of enzyme concentration, temperature, pH, and inhibitors upon enzyme activity.

Discussion

In biological systems, reactions are catalyzed by enzymes, which speed up reactions while operating at mild temperature and pH. Like all catalysts, enzymes lower the energy of activation needed for a reaction to take place. For example, an enzyme in blood called carbonic anhydrase converts carbon dioxide and water to carbonic acid. The enzyme catalyzes the reaction of about 35 million molecules of CO_2 every minute. See Figure 38.1

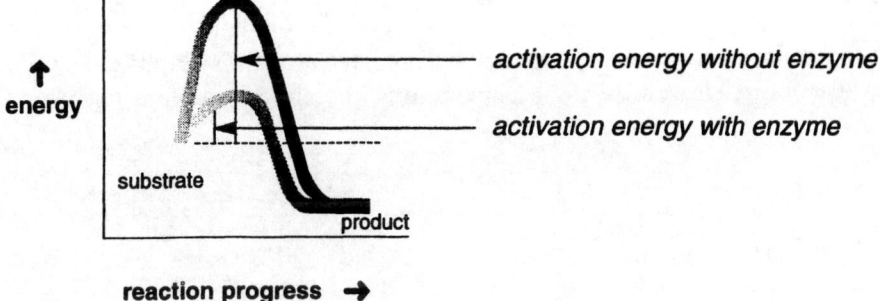

Figure 38.1 Activation energy of a reaction with and without an enzyme (catalyst)

In this experiment you will use amylase, an enzyme that begins the hydrolysis of carbohydrates (amylose) in the mouth. In the presence of amylase, a sample of starch will undergo hydrolysis to give smaller polysaccharides, dextrins, maltose, and glucose.

$$\text{Starch (amylose)} \xrightarrow{\textit{Amylase}} \text{Smaller polysaccharides, dextrins, maltose} \xrightarrow{\textit{Amylase}} \text{Glucose}$$

Testing Enzyme Activity

We will determine the reaction of amylase with starch by testing for starch using iodine reagent. When iodine is added to a starch solution, a blue-black color is produced. However, if amylase is present and the starch is hydrolyzed, the iodine test is no longer positive. Only the red or gold color of the iodine solution is seen. The faster the amylase breaks down starch, the more quickly the blue-black color is lost. If the blue-black color persists, the enzyme is inactive.

The hydrolyzed product glucose can be detected by using Benedict's reagent. With glucose, the Benedict's reagent turns from blue to a green or reddish-orange with higher glucose concentrations. Table 38.1 summarizes the tests for amylose and glucose.

Table 38.1 *Detection Tests for Starch and Glucose*

Starch (Amylose)	Glucose
Positive iodine test (Turns deep blue with iodine)	Negative iodine test (No color change with iodine. Remains yellow-orange)
Negative Benedict's test (Remains blue)	Positive Benedict's test (Turns green to reddish-orange)

A. Effect of Enzyme Concentration

During catalysis, an enzyme combines with the reactant or *substrate* of a reaction to give an *enzyme-substrate* complex. To form this complex, the substrate fits into the *active site,* where reaction takes place. The *products* are released and the enzyme is ready to catalyze another reaction.

$$E \ + \ S \ \underset{\longleftarrow}{\overset{\longrightarrow}{}} \ ES \ \underset{\longleftarrow}{\overset{\longrightarrow}{}} \ E \ + \ P$$

Enzyme *Substrate* *Enzyme–substrate complex* *Enzyme* *Product*

If the enzyme concentration is increased while substrate concentration is constant, the rate of the reaction will increase. With more enzyme, more substrate molecules can react.

B. Effect of Temperature

The *optimum* temperature is the temperature at which an enzyme operates at maximum efficiency. At low temperatures, the rate of reaction is slowed. At high temperatures, the enzyme protein is denatured. See Figure 38.2.

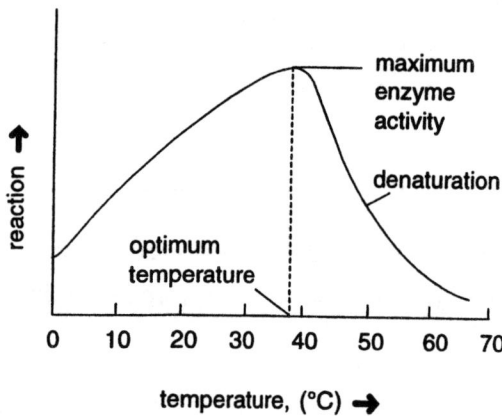

Figure 38.2 Effect of temperature on enzyme activity

C. Effect of pH

At the optimum pH, an enzyme is most active. At pH values above and below optimum, the protein structure of the enzyme is altered, which can severely reduce the enzyme's activity. See Figure 38.3.

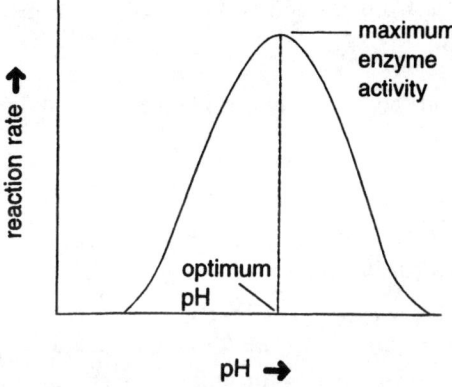

Figure 38.3 Effect of pH on enzyme activity

D. Inhibition of Enzyme Activity

Substances that limit or stop the catalyzing activity of an enzyme are called *inhibitors*. A *competitive inhibitor* blocks the active site of an enzyme, while a *noncompetitive inhibitor* binds to the surface of the enzyme and disrupts the structure of the active site. An *irreversible inhibitor* forms bonds with side-chains of the amino acids in the active site, which makes the enzyme inactive.

Lab Information

Time: 2½ hr
Comments: Tear out the report sheets and place them beside the matching procedures.

Related topics: Enzymes, active site, lock-and-key theory, enzyme activity, factors affecting enzyme activity, inhibition, denaturation

Experimental Procedures

WEAR YOUR SAFETY GOGGLES!

Constant temperature water baths: If commercial water baths are not available, prepare your own. Fill three 250-mL or 400-mL beakers about half full of tap water. Warm one to about 37°C, and heat one to boiling. To the third, add ice to give an ice bath with a low temperature (0-5°C)

Type of Bath	Temperature of Water Baths
Boiling hot water bath	100°C
Warm water	37°C (body-temperature)
Ice bath	0° (or lower than 5°C)

Starch solution: Place 50 mL of 1% starch solution in a small beaker.

Iodine reagent: Obtain a small amount of iodine reagent in a second small beaker. Dropper bottles of iodine may be available. In a third small beaker, place 40 mL of a freshly prepared amylase solution. Obtain a spot plate or plastic sheet for testing. Use clean droppers and rinse. Rinsing the droppers is important to avoid transferring enzyme to other samples.

Amylase preparation: Your instructor may provide a commercial amylase solution.

Reference Tests

Starch: Place a few drops of starch solution in a depression in the spot plate. Add 1 drop of iodine reagent. The reaction with starch should give a deep blue-black color.

Glucose sample: Place 3 mL of 1% glucose in a test tube. Add 2 mL of Benedict's solution and heat for 3–4 minutes in a boiling water bath. The reaction should give a solid with a red-orange color.

Visual Color Reference

As you proceed with each experiment, you will check enzyme activity by adding iodine to the starch mixture with iodine. When enzyme activity is high, the time required for the starch to hydrolyze will be very short. When the enzyme is slowed down or inactive, the blue-black color will be seen for a longer time. By observing the disappearance of starch, you can assess the relative amount of enzyme activity as follows:

Iodine Test for Starch	Amount of Starch Remaining	Enzyme Activity Level	
Dark blue-black	All	None	0
Blue	Most	Low	1
Light brown	Some	Moderate	2
Gold	None	High	3

A. Effect of Enzyme Concentration

Materials: Test tubes, test tube rack, thermometer, 37°C water bath, droppers, spot plate (or plastic sheet), amylase preparation, 1% starch (buffered pH 7.0), iodine reagent, Benedict's solution, boiling water bath, 5- or 10-mL graduated cylinder

A.1 Fill a large beaker about 2/3 full of tap water. Warm to about 37°C and try to maintain the temperature. Your lab may have a commercial water bath set at 37°C.

Place 4 mL of 1% starch in each of four separate test tubes labeled 1–4. Place 4 mL of amylase solution in a fifth tube. Place all the tubes in a 37°C water bath for 5 minutes. Tube 1 without enzyme is the control. Add 3 drops of the warmed amylase solution to the second tube, 6 drops to the third tube, and 10 drops to the fourth tube. Mix quickly and return the test tubes immediately to the 37°C water bath. Record the time at which you add enzyme. Keep the temperature of the water bath close to 37°C, adding water as needed.

Test Tube	Amylase Solution
1	0 (control)
2	3 drops
3	6 drops
4	10 drops

Immediately, transfer four drops of each reaction mixture (use clean droppers) to a spot plate (or plastic sheet). Add 1 drop of iodine reagent to each. Record your observations for each sample. Use the visual color reference to assess the enzyme activity. Rinse the spot plate or use a new section of plastic sheet. Repeat the test again in 5 and 10 minutes.

A.2 (*Optional*) In the test tubes where hydrolysis occurred, the presence of glucose can be confirmed using Benedict's test. Add 3 mL of Benedict's solution to each test tube and place them in a boiling water bath for 3–4 minutes. The appearance of a green to orange-rust color indicates the presence of glucose. If the solution remains clear blue, no hydrolysis of the starch has taken place.

A.3 Make a graph comparing the enzyme activity and the amount of amylase solution at 10 minutes.

B. Effect of Temperature

Materials: Test tubes, test tube rack, test tube holder, droppers, 5- or 10-mL graduated cylinder, beakers for water baths: ice (0°C), warm (37°C), and boiling (100°C), amylase preparation, spot plate (or plastic sheet), 1% starch, iodine reagent

B.1 Place 4 mL of 1% starch solution in each of three test tubes. Place one tube in a boiling water bath, one tube in the 37°C water bath, and one tube in the ice bath. Add 4 mL of amylase solution to three other test tubes and place one each in the three water baths. Let them remain in the water baths for about 10 minutes to allow the solutions to reach the bath temperature.

For each temperature bath, proceed as follows:
a. Record the temperature of the bath. Remove the test tubes, pour the contents together in one tube, mix, and return the mixture to the same temperature bath.

b. After 10 minutes transfer four drops of the mixture to a spot plate (or plastic sheet). Add 1 drop of iodine to each sample. Record the color and activity level.

B.2 Make a graph of the enzyme activity level and the temperature.

C. Effect of pH

Materials: Test tubes, test tube rack, amylase preparation, 37°C water bath, buffers (pH 2, 4, 7, 10), spot plate (or plastic sheet), 1% starch, iodine reagent

C.1 Place 4 mL of buffer solutions of pH 2, 4, 7, and 10 in separate test tubes. Add 4 mL of amylase solution in each. In four other test tubes place 4 mL of 1% starch solution. Place all of the test tubes in a 37°C water bath for about 5 minutes. Pour each of the 1% starch solutions into a separate buffer–amylase tube. Mix and return the mixtures to the 37°C water bath.

After 15 minutes, remove four drops of each reaction mixture (use clean droppers each time) and place in the spot plate or on the plastic sheet. Add 1 drop of iodine reagent to each. Record your observations and the enzyme activity level for each reaction mixture.

C.2 Make a graph of the enzyme activity level and pH.

D. Inhibition of Enzyme Activity

Materials: Test tubes (6), test tube rack, 5- or 10-mL graduated cylinder, amylase preparation, 37°C water bath, solutions in dropper bottles: 1% NaCl, 1% $AgNO_3$, 1% $Pb(NO_3)_2$, 1% $HgCl_2$, 95% ethanol, 1% starch, iodine reagent, spot plate (or plastic sheet)

In one test tube, place 4 mL of amylase solution and 10 drops of 1% NaCl. In a second test tube, place 4 mL of amylase solution and 10 drops of 95% ethanol. In a third test tube, place 4 mL of amylase solution and 10 drops of one of the inhibitors that your instructor assigns: 1% $AgNO_3$, 1% $Pb(NO_3)_2$, or 1% $HgCl_2$.

Place 4 mL of 1% starch in three other test tubes. Place all six tubes in a 37°C water bath for 5 minutes. Then combine the starch solutions with the amylase solutions, mix, and return the tubes to the 37°C bath for 15 minutes.

After 15 minutes, remove four drops of each reaction mixture (use clean droppers each time) and place in the spot plate or on the plastic sheet. Add 1 drop of iodine reagent to each. Record your observations and the enzyme activity level for each of the reaction mixtures.

Report Sheet - Lab 38

Date _____   Name _____

Section _____   Team _____

Instructor _____

Pre-Lab Study Questions

1. What is the substrate of the enzyme amylase?

2. What are the products of amylase action?

3. What happens to enzymes at high temperatures?

4. Reference tests:

   Color of iodine test for starch   _____

   Color of Benedict's test for glucose   _____

Report Sheet - Lab 38

A.   Effect of Enzyme Concentration

A.1 Time	Color Produced by Iodine Test and Activity Level							
	No amylase		3 drops amylase		6 drops amylase		10 drops amylase	
0 minutes								
5 minutes								
10 minutes								

	Color Produced by Benedict's Test				
A.2 Benedict's test					

A.3   Graph: Enzyme activity at 10 minutes vs. enzyme concentration

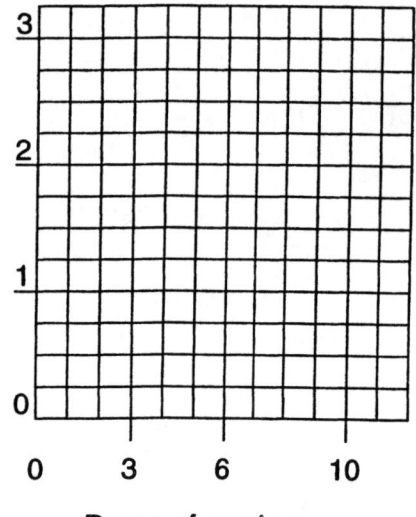

Enzyme activity level

Drops of amylase

Questions and Problems

Q.1 At which enzyme concentration was starch hydrolyzed the fastest? Slowest?

Q.2 Describe the effect of enzyme concentration on enzyme activity.

Report Sheet - Lab 38

B. Effect of Temperature

B.1

Color Produced by Iodine Test and Activity Level					
0°C	Level	37°C	Level	100°C	Level

B.2 Graph: Enzyme activity level and temperature

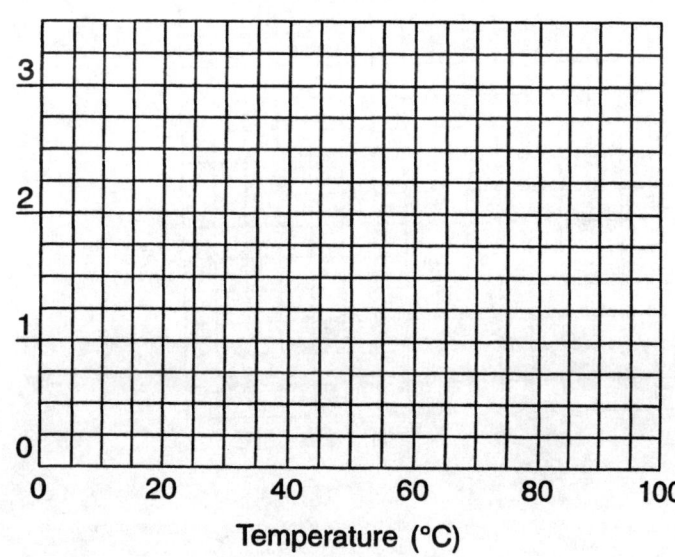

Enzyme activity level

Temperature (°C)

Questions and Problems

Q.3 What was the optimal temperature for amylase?

Q.4 Why did the enzyme activity differ at 0°C and at 100°C?

Report Sheet - Lab 38

C. Effect of pH

C.1   Color Produced by Iodine Test and Activity Level

C.1	Color Produced by Iodine Test and Activity Level							
	pH 2	Level	pH 4	Level	pH 7	Level	pH 10	Level

C.2   Graph: Enzyme activity level and pH

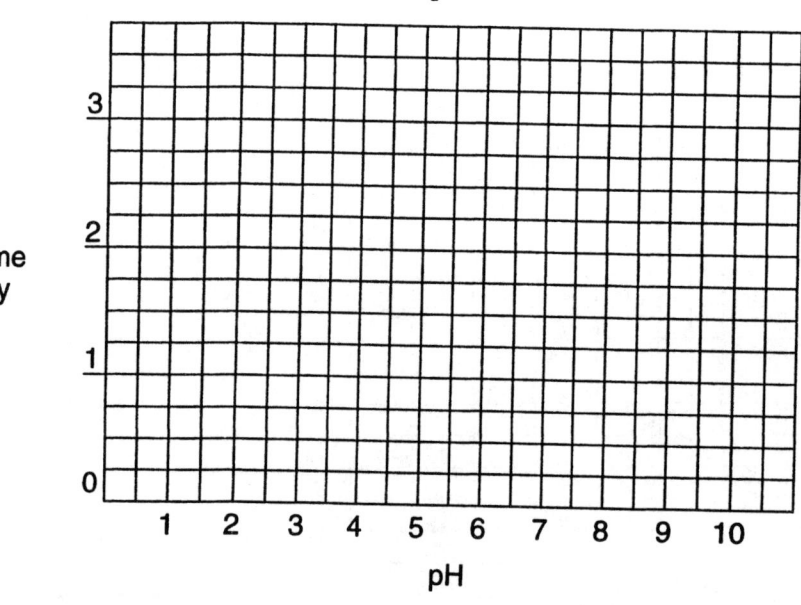

Questions and Problems

Q.5 How is amylase affected by a low pH? By a high pH? Explain.

Q.6 What was the optimum pH for amylase?

Q.7 a.   During digestion, the pH in the stomach is 2. What does this indicate about the optimum pH of pepsin, an enzyme that hydrolyzes protein in the stomach?

b.   What happens to the activity of pepsin when it enters the small intestine where the pH is 8?

Report Sheet - Lab 38

D. Inhibition of Enzyme Activity

Color Produced by Iodine Test and Activity Level					
NaCl	Level	Ethanol	Level	Inhibitor _____	Level

Questions and Problems

Q.8 In which reaction mixture(s) did hydrolysis of starch occur?

Q.9 What substances added to the mixture were inhibitors?

Q.10 How might those substances inhibit enzyme action?

Q.11 What are some differences and/or similarities in the type of inhibition caused by heat, acid or base, and heavy metal ions on enzyme activity?

B. Standardization of Vitamin C

Materials: Vitamin C (100-mg tablet), mortar and pestle, 250-mL Erlenmeyer flask, 50-mL graduated cylinder, 50-mL buret, buret clamp, small funnel, small beaker, 0.1 M HAc, 1% starch, iodine solution, dropper

B.1 Obtain a vitamin C tablet (100-mg vitamin C). If not 100 mg, record the amount (mg) of vitamin C in the tablet as stated on the label. Crush the tablet and transfer it to a 250-mL Erlenmeyer flask. Add 50 mL of distilled water, 2 mL of 0.1 M HAc (acetic acid), and mix. Add 10 drops of 1% starch solution.

B.2 Place a 50-mL buret in a buret clamp on the ring stand. Carefully fill the buret just above the zero mark with the iodine solution. Drain the iodine down to the zero mark. Record the initial level of the iodine solution. See Figure 39.1.

Caution: Keep iodine reagent away from clothes and skin.

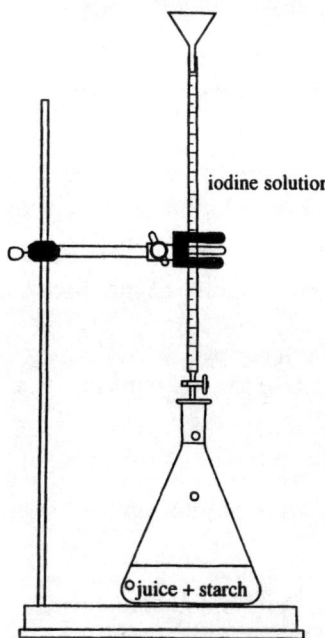

iodine solution

juice + starch

Figure 39.1 Titration setup for vitamin C analysis

Begin adding the iodine solution from the buret to the vitamin C solution until the solution has a deep-blue color. This endpoint is reached after all of the vitamin C has been oxidized and the next drop of iodine solution is not reduced. Record the final reading of the iodine solution in the buret. Calculate the volume of iodine solution used in the titration.

B.3 Calculate the mass (mg) of vitamin C that reacts with 1 mL of iodine solution.

$$\frac{\text{Mass (mg) of vitamin C in tablet}}{\text{Volume (mL) of iodine solution}} = \text{mg vitamin C oxidized by 1 mL iodine solution}$$

C. Analysis of Vitamin C in Fruit Juices and Fruit Drinks

Materials: 125-mL Erlenmeyer flasks, funnel, 50-mL buret, buret clamp, small funnel, small beaker, cheesecloth, 50-mL graduated cylinder, 0.1 M HAc (acetic acid), 1% starch indicator, iodine solution
Fruit juices/drinks: orange, grapefruit, drinks such as HiC or Snapple, or powdered drinks such as Tang or Kool-Aid, or vegetables pureed in a blender or juicer

The experiment works best if the juice is colorless or has a light color.

C.1 Obtain a sample of fruit juice or fruit drink for vitamin C analysis. Record the type of juice or drink. If the juice has a lot of pulp (fiber), pour the juice through cheesecloth that covers a funnel. If suspension particles go through, use two layers of cheesecloth. Use a pipet or small graduated cylinder to transfer a 20-mL sample of the juice into a clean 125-mL Erlenmeyer flask. Add 30 mL of distilled water, 2 mL of 0.1 M HAc (acetic acid), and 10 drops of 1% starch indicator.

Powered drink: If you have a powdered fruit drink such as Tang or Kool-Aid, weigh 1.0 g of the powder and place in a 125-mL Erlenmeyer flask. Add 50 mL of water to the powder, 2 mL of 0.1 M HAc, and 10 drops of 1% starch indicator, and mix. The sample is ready to titrate with iodine.

Vegetables: Weigh 10.0 g of a vegetable. Puree the vegetable in a blender or juicer with 20 mL of water. Filter the puree through cheesecloth. Add water to give a total volume of 50 mL. Add 2 mL 0.1 M HAc and 10 drops of 1% starch indicator.

C.2 Record the initial reading of the level of iodine solution in the buret. Place the flask with the juice and starch mixture under the buret and begin adding iodine solution until the indicator just turns a dark blue. Record the final buret reading for the level of iodine solution.

C.3 Calculate the volume of iodine solution used to reach the endpoint of the titration.

C.4 Calculate the mg of vitamin C in the fruit juice sample. Use the value of mg of vitamin C per 1 mL of iodine solution obtained in step B.3.

$$\_\_\_ \text{ mL iodine solution} \times \frac{\text{mg vitamin C}}{1 \text{ mL iodine solution}} = \text{mg vitamin C}$$

Repeat the titration with another sample of juice or other vitamin C sample.

D. Heat Destruction of Vitamin C

Materials: 250-mL Erlenmeyer flasks, funnel, cheesecloth, 50-mL buret, buret clamp, small funnel, small beaker, 50-mL graduated cylinder, 0.1 M HAc (acetic acid), 1% starch indicator, iodine solution, Bunsen burner or hot plate, ice-water bath
Fruit juices/drinks: orange, grapefruit, drinks such as HiC or Snapple, or powdered drinks such as Tang or Kool-Aid, or vegetables pureed in a blender or juicer

D.1 Place 20 mL of a juice in part C that had a high vitamin C content in each of two 250-mL Erlenmeyer flasks. Add 50 mL of water to each. Boil one sample for 10 minutes, and the other for 30 minutes. After 10 minutes, remove the first flask and place it in an ice-water bath. Add 2 mL of 0.1 M HAc and 10 drops of starch indicator. Fill a buret with iodine solution and record the initial level. Titrate with iodine solution to the deep-blue endpoint. Record the final level of iodine solution. Repeat with the other sample.

D.2 Calculate the volume of iodine solution used in each titration.

D.3 Calculate the mg of vitamin C present in each of the heated samples.

D.4 Using the value in C.4 of the mg of vitamin C in the juice sample, calculate the mg of vitamin C that were destroyed by heating the juice.

Report Sheet - Lab 39

Date _____   Name _____

Section _____   Team _____

Instructor _____

Pre-Lab Study Questions

1. What is the difference between a water-soluble vitamin and a fat-soluble one?

2. What is the metabolic role of vitamin C?

3. What foods contain large quantities of vitamin C?

4. What disease is associated with a diet lacking in vitamin C?

Report Sheet - Lab 39

A. Solubility of Vitamins

Vitamin	Soluble in Water	Soluble in CH_2Cl_2	Water or Fat Soluble?	Metabolic Function

Questions and Problems

Q.1 Which vitamins are water soluble?

Q.2 Which vitamins are fat soluble?

Q.3 Which vitamins are required?

Report Sheet - Lab 39

B. Standardization of Vitamin C

B.1 Mass of vitamin C (from label) _____mg

B.2 Initial buret reading _____

Final buret reading _____

Volume of iodine solution used _____

B.3 mg Vitamin C per 1 mL iodine solution _____ mg /1 mL iodine solution

(*Show calculations.*)

C. Analysis of Vitamin C in Fruit Juices and Fruit Drinks

C.1 Type of juice or drink _____ _____

C.2 Initial buret reading _____ _____

Final buret reading _____ _____

C.3 Volume of iodine solution used _____ _____

C.4 mg Vitamin C in the juice sample _____ _____

(*Show calculations.*)

Questions and Problems

Q.4 Which of the juices that you analyzed had the most vitamin C?

Q.5 If the daily requirement for vitamin C is 75 mg, how many milliliters (or grams) of each sample would you need to consume to meet the minimum daily requirement?

Report Sheet - Lab 39

D.   Heat Destruction of Vitamin C

Sample used _____

	Boiled 10 Minutes	**Boiled 30 Minutes**
D.1 Initial buret reading	_____	_____
Final buret reading	_____	_____
D.2 Volume of iodine solution used	_____	_____
D.3 mg Vitamin C in heated sample (*Show calculations.*)	_____mg	_____mg
D.4 mg Vitamin C destroyed	_____mg	_____mg

Questions and Problems

Q.6   Does heating affect the vitamin C content of a fruit juice?

Q.7   If vitamin C tablets are stored in a warm, humid bathroom cabinet, what might happen to the vitamin C content after a while?

Q.8   If you wish to keep most of the vitamin C content of your vegetables, how should you prepare them for dinner?

Lab Information

Time: 2-3 hr
Comments: Tear out the report sheets and place them beside the matching procedures.
Related topics: Deoxyribonucleic acid, nitrogenous bases, deoxyribose, nucleosides, nucleotides,
                 replication

Experimental Procedures

WEAR YOUR PROTECTIVE GOGGLES!

A. Components of DNA

Materials: organic model kits

A.1 Use the C, H, N, and O atoms in a model kit to make a model of one of the purines in DNA nucleotides. Have another lab team make a model of the other purine. Draw the structures of the purines found in the nucleotides of DNA. Save these models.

A.2 Use the C, H, and O atoms in a model kit to make a model of deoxyribose. Draw the structure of deoxyribose. Save this model.

A.3 Use the P, H and O atoms in a model kit to make a model of phosphate. Attach the phosphate group and the purine you constructed to the deoxyribose sugar. Draw the structure of this nucleotide. Write its name. Save this model.

A.4 With your neighbor lab team, combine two nucleotides to form a dinucleotide. Draw the structure of this dinucleotide. Write its name. Combined the two nucleotides to make a different nucleotide. Draw the structure of this dinucleotide.

B. Extraction of DNA

Different student teams may extract DNA from different DNA sources and compare results or the procedure may be run initially with onions and then repeated using a DNA source selected by each student team.

Materials: DNA sources: white onions, variety of other DNA plant sources such as cauliflower, broccoli, garlic, split peas, bananas, and/or animal sources such as chicken liver, calf thymus.
Lab items: knife, two 250-mL beakers, thermometer, hot water bath (400-mL beaker about $^1/_2$ full of water, iron ring, wire screen, and Bunsen burner), blender, cheesecloth or filter paper, microscope
Lab chemicals SDS-NaCl-EDTA buffer solution (Your instructor will prepare this buffer: 50 g sodium dodecyl sulfate, 50 g NaCl, 5 g sodium citrate, 0.2 mL of 0.5 MEDTA and water to make 1 L), isopropanol (placed in an ice bath to keep cold), citrate buffer (0.15 M NaCl, 0.015 M sodium citrate)

1. Obtain about 50 g of a white onion or other DNA source. Use a knife to dice the onion or other DNA source into small pieces. Place the pieces in a 250-mL beaker and add 50 mL of the extraction (SDS-NaCl-EDTA) buffer.

2. Prepare a hot water bath. Place the beaker in the hot water bath, and heat to 60°C. Maintain a temperature of about 60°C by adjusting or removing the flame of the heat source. Allow the beaker to remain in water at 60°C for 15 minutes. (Any longer time in hot water will start to break down DNA.) Stir the mixture occasionally. Remove the beaker and place it in an ice bath for 10 minutes.

Pour the cooled mixture into a blender and blend the contents for 60 seconds using 15-second bursts. Pour the mixture through two pieces of cheesecloth placed over a clean 250-mL beaker. This filtering may take as long as one hour. You may need to use new sets of cheesecloth as it becomes clogged with cell debris.

3. Measure the volume of the filtered onion liquid collected in the beaker and obtain an equal volume of ice-cold isopropanol. If the alcohol is not ice cold, cool it first in an ice bath. Holding the beaker at an angle, slowly pour the alcohol down the side. Allow the solution to sit for 2 minutes. An alcohol layer should form on the top of the filtrate. Because DNA is not soluble in alcohol, the whitish, viscous strands of DNA should precipitate out of the alcohol layer and form a whitish interface. (Other components of the mixture will remain soluble in the alcohol layer.)

Place a glass rod into the solution and turn it slowly to wind the DNA threads into a ball on the end of the glass rod. When you remove the glass rod the DNA will look like a viscous blob. Transfer the DNA from the glass rod to a paper towel and let it dry.

4. Observe the DNA obtained by other student teams with different DNA sources, or repeat the procedure above with another DNA source.

B.1 Observe texture and physical properties of the DNA you have extracted.

B.2 Examine the DNA under a microscope. Record your observations.

Report Sheet - Lab 40

Date _____ Name _____

Section _____ Team _____

Instructor _____

Pre-Lab Study Questions

1. Why does DNA have a slight negative charge?

2. What would homogenization and heating do to cell membranes?

A. Components of DNA

Structures of Models (Lab Team 1)	Structures of Models (Lab Team 2)
A.1	A.1
A.2	A.2
A.3	A.3

(Continued)

Report Sheet - Lab 40

Structures of Models (Lab Team 1)	Structures of Models (Lab Team 2)
A.4	A.4

B. Extraction of DNA

DNA Source B.1	Texture and Appearance	B.2 Appearance under a microscope

Report Sheet - Lab 40

Questions and Problems

Q.1 Why is it necessary to heat the DNA source and buffer mixture?

Q.2 Why was an alcohol added to the filtered onion liquid?

Q.3 What similarities did you observe in the appearance of DNA from different DNA sources? Why?

Q.4 How would be DNA extracted from an onion be different from the DNA extracted from cauliflower? Why?

Q.5 Write the complementary base sequence for the following segment of DNA.

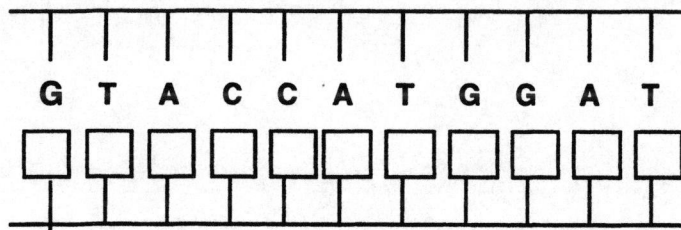

G T A C C A T G G A T

41

Digestion of Foods

Goals

- Identify the types of hydrolysis reactions that take place during the digestion of food.
- Use chemical tests to identify the hydrolysis products of carbohydrates, fats, and proteins.

Discussion

The digestive processes utilize enzymes to carry out the hydrolysis of large food molecules to molecules small enough to dialyze through the intestinal wall into the blood or lymph. See Figure 41.1.

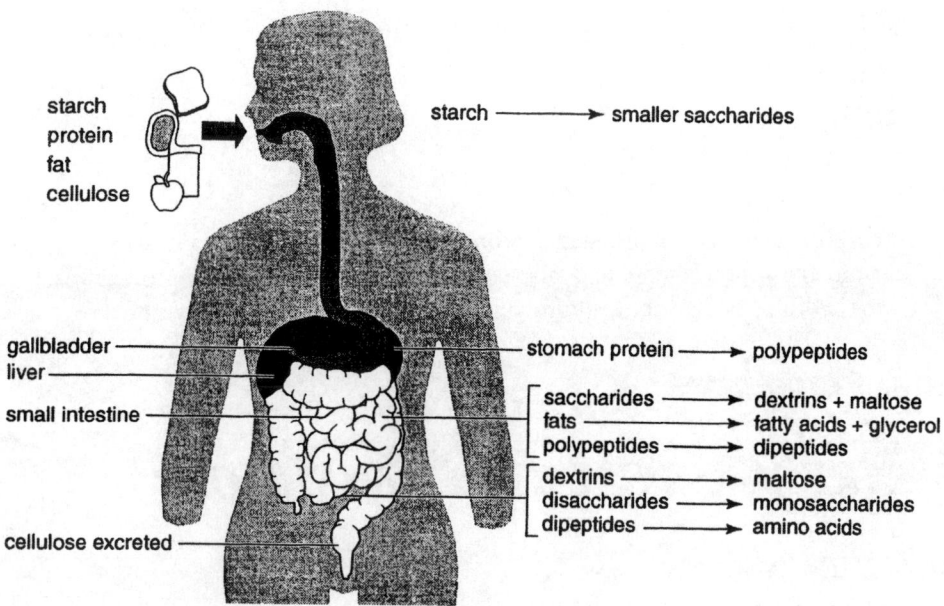

Figure 41.1 Sites of digestion in the human body

A. Digestion of Carbohydrates

Starch, a major carbohydrate in our foods, provides about 50% of our caloric intake. In order to use starch, it must be hydrolyzed into glucose molecules. Digestion of starch begins in the mouth by the action of an enzyme, salivary amylase. Hydrolysis continues in the small intestine through the action of pancreatic amylase, and maltase.

$$\text{Starch (amylose)} \xrightarrow{\;Amylase\;} \text{maltose} \xrightarrow{\;Maltase\;} \text{glucose}$$

B. Digestion of Fats

Approximately 25–30% of our diet consists of lipids, primarily fats (triglycerides). Chemically, fat is an ester of glycerol and fatty acids. Digestion of fats begins in the intestine with bile salts and the enzymatic action of lipases obtained from the gallbladder. The bile salts cause the fat to break up into smaller droplets (emulsification), increasing the surface area, and the lipases hydrolyze the ester bonds of the fats.

$$\text{Fats} \xrightarrow{\text{\textit{Pancreatic lipase}}} \text{glycerol} + \text{fatty acids}$$

C. Protein Digestion

Proteins, which make up about 20–25% of our diet, begin to be digested in the stomach where HCl activates the proteases such as pepsin that begin the hydrolysis of peptide bonds. Other enzymes continue to hydrolyze polypeptides and dipeptides.

$$\text{Proteins} \xrightarrow{\text{\textit{Pepsin, chymotrypsin}}} \text{peptides, dipeptides} \xrightarrow{\text{\textit{Dipeptidases}}} \text{amino acids}$$

Lab Information

Time: 2 hr
Comments: Be careful when you work with boiling water.
                   Tear out the report sheets and place them beside the matching procedures.
Related topics: Carbohydrate, fat, protein, hydrolysis, digestion, digestive enzymes

Experimental Procedures

ARE YOUR SAFETY GLASSES ON?

A. Digestion of Carbohydrates

Materials: Test tubes (2), test tube rack, 10-mL graduated cylinder, droppers, spot plates (or plastic sheets), boiling water bath (large beaker), hot plate or Bunsen burner, amylase solution, 1% starch solution, iodine solution, Benedict's solution

A.1   Hydrolysis of Starch

Amylase preparation: Obtain a spot plate or plastic sheet to test each mixture for starch. In two test tubes, place 4 mL of 1% starch solution. To one, add 10 drops of amylase solution; to the other, add 10 drops of water. Mix thoroughly.

After 5 minutes, use clean droppers to transfer four drops of each mixture to a spot plate (or plastic sheet). Add 1 drop of iodine solution to each. A blue-black color indicates that starch remains in the mixture. Repeat the tests at 10 minutes. Record your observations. Save the mixtures for part A.2.

A.2 Test for Glucose

The presence of glucose as a final product of starch digestion can be determined by adding 3 mL of Benedict's reagent to each of the two test tubes from part A.1. Place the test tubes in a boiling water bath for 5 minutes. Record the colors that form. Benedict's test is positive for glucose if the orange color of Cu_2O forms. Record your observations.

$$Glucose \ + \ 2Cu^{2+} \longrightarrow \ Gluconic \ acid \ + \ Cu_2O(s)$$

Blue   Reddish-orange

B. Digestion of Fats

Materials: Test tubes, test tube rack, 50-mL buret, buret clamp, small funnel, small beaker, 25- or 50-mL graduated cylinder, two 250-mL Erlenmeyer flasks, droppers, 37°C water bath, safflower oil, bile salt solution, whole milk, 2% pancreatin, 0.1 M NaOH, pH meter or pH paper, phenolphthalein

B.1 **Action of bile salts** Place 20 drops of safflower oil in each of two test tubes. Add 4 mL water to one test tube. To the other sample, add 2 mL water and 2 mL of bile salt solution. Mix thoroughly. Record your observations of the mixtures in the test tubes. Stir occasionally. After 15 minutes, observe the mixtures in the test tubes again. Look for the separation of layers, or emulsification by the bile salts. Record your observations.

B.2 **Hydrolysis by lipase** Set up a buret containing 0.1 M NaOH. Place 15 mL of whole milk in each of two 250-mL Erlenmeyer flasks. Add 10 mL of 2% pancreatin to each flask and mix thoroughly. Place one flask in a 37°C water bath.

Determine the pH of the milk sample and then titrate the milk sample. Add 3–5 drops of phenolphthalein. Add the 0.1 M NaOH from the buret until a permanent light pink color is obtained. This marks the endpoint. Record the number of milliliters of NaOH used to reach the endpoint.

After 60 minutes, remove the other milk sample from the 37°C water bath. Determine the pH of the sample. Add 3–4 drops of phenolphthalein and titrate with 0.1 M NaOH. Record the number of milliliters of NaOH used to reach the light, permanent pink endpoint.

C. Protein Digestion

Materials: Test tubes (3), test tube rack, 10-mL graduated cylinder, dropper, hard-boiled egg white, 2% pepsin, 0.1 M HCl, 37°C water bath

Cut a piece of egg white into three small pieces about 1 cm in length. Add one piece of egg white to each of three test tubes with the following solutions:

Test Tube	Solutions		
1	4 mL water	+	1 mL (20 drops) 0.1 M HCl
2	4 mL of 2% pepsin	+	1 mL (20 drops) 0.1 M HCl
3	4 mL of 2% pepsin	+	1 mL (20 drops) water

Place the test tubes in a 37°C water bath. Record the appearance of the egg white initially. After 30 minutes, record changes in the egg white in each test tube. Return the test tubes to the water bath and observe again after an additional 30 minutes.

Report Sheet - Lab 41

Date _____   Name _____

Section _____   Team _____

Instructor _____

Pre-Lab Study Questions

1. What type of compound and bond is hydrolyzed by the following?

    a. α-amylase

    b. lipase

2. Why are the reactions of digestion called hydrolysis reactions?

A. Digestion of Carbohydrates

A.1 Hydrolysis of starch

Time (minutes)	Starch	Starch + Amylase
5		
10		

A.2 Test for glucose

Benedict's solution		
Is glucose present?		

Questions and Problems

Q.1 What is (are) the final product(s) of carbohydrate digestion?

Q.2 Why do some people need lactase when they consume milk products?

Report Sheet - Lab 41

B. Digestion of Fats

B.1 Action of bile salts

	Oil Only	Oil + Bile Salts
Initially		
15 minutes		

B.2 Hydrolysis by lipase

Time (minutes)	pH	Volume of 0.1 M NaOH
0		
60		

Questions and Problems

Q.3 What is the function of bile salts in digestion? What organ produces bile salts?

Q.4 Which products of fat hydrolysis would cause a change in pH of the whole milk sample?

Q.5 Write the equation for the action of pancreatin (lipase) on a molecule of tristearin.

Report Sheet - Lab 41

C. Protein Digestion

Appearance of Egg White	Water + HCl	Pepsin + HCl	Pepsin + H_2O
Initial			
After 30 minutes			
After 60 minutes			

Questions and Problems

Q.6 Why does a person with a low production of stomach HCl have difficulty with protein digestion?

Q.7 What are the sites and enzymes that digest protein in the body?

Q.8 What are the final products of protein digestion?

Goals

- Use tests to determine pH, specific gravity, and the presence of electrolytes and organic compounds in urine-like specimens.

- Use chemical tests to analyze for the presence of proteins, glucose, and ketone bodies in urine-like specimens.

Discussion

A. Color, pH, and Specific Gravity

Examining a urine specimen can give diagnostic information about the processes occurring within the body. The pH, the amounts of electrolytes, uric acid, and glucose can all lead to conclusions about the functioning of the kidneys and liver and the general state of health of the individual. Typically, a person excretes about 1 liter of urine daily. This volume varies with the amount of liquid intake, the temperature, exercise, and the use of substances such as caffeine.

Urine usually has a pH around 6.0, although this varies considerably with diet and activity and can range from 5 to 9 at different times. Urine normally has a light yellow color derived from the breakdown of bilirubin formed during the destruction of red blood cells. The normal range for specific gravity is 1.005 to 1.030. In this experiment, you will test the pH of a urine-like sample, measure its specific gravity, and note its color.

B. Electrolytes

Urine is about 96% water. The other 4% consists of waste products being eliminated from the body to maintain proper osmotic pressure, electrolyte levels, and pH. Urine normally contains the inorganic ions Cl^-, HCO_3^-, SO_4^{2-}, PO_4^{3-}, K^+, Na^+, NH_4^+, and Ca^{2+}. The presence of sodium ion can be determined by a flame test. The chloride ion will be detected using silver nitrate. Barium ion will be used to detect the presence of sulfate ion. Phosphate ion will be identified with ammonium molybdate.

C. Glucose

Glucose is not normally detected in the urine although the kidneys do excrete very small amounts of glucose. However, when the glucose level in the blood exceeds the renal threshold, glucose may show up in urine (glucosuria). Concentrations as low as 0.1 g/dL could be abnormal and indicate diabetes mellitus or liver damage. In severe diabetes, glucose levels may reach 5–10 g/dL urine. Reagent strips such as Clinitest or Multistix are used at home and in hospitals to determine the glucose levels in urine. On the strip, glucose oxidase converts glucose to gluconic acid and hydrogen peroxide. A peroxidase enzyme catalyzes the reaction of the hydrogen peroxide with a potassium iodine complex to give colors that range from green to brown.

D. Ketone Bodies

Ketone bodies such as acetone and acetoacetic acid are normally not present in urine. However, they do appear in urine in cases of low-carbohydrate diets, starvation, diabetes mellitus, and liver damage. Reagent test strips such as Ketostix or Multistix are used to detect ketone bodies. The test is positive when acetoacetic acid or acetone in the urine reacts with nitroprusside on the strip. The reaction produces a purple color that deepens in intensity with larger amounts of the ketone bodies.

E. Protein

Normally no protein is detected in urine, although the kidney excretes very small amounts. Typically a person excretes 50–100 mg protein in one day. However, in *proteinuria,* urinary protein levels higher than normal may indicate disease or damage to the kidneys or urinary tract. Protein can be detected by heating a portion of the urine specimen to coagulate the protein. Albustix or Multistix test strips can be used to detect protein in urine. The development of a yellow-green to blue-green color is a positive reaction for protein.

F. Urobilinogen

Urobilinogen is a bile-related pigment that is excreted at low levels (0.1 to 1 Ehrlich units/dL urine). A reagent strip produces a range of brownish-orange colors when the urobilinogen level is elevated (2 Ehrlich units/dL or greater).

Lab Information

Time: 2 hr

Comments: Be sure to read the test strips at the indicated time after immersing the strip in the fluid.Tear out the report sheets and place them beside the matching procedures.

Related topics: Electrolytes, pH, ketone bodies

Experimental Procedures

WEAR YOUR SAFETY GOGGLES!

A. Color, pH, and Specific Gravity

Materials: "Urine" specimens (normal and abnormal, prepared by instructor), pH paper or pH meter, urinometer (hydrometer) *or* a small graduated cylinder and beaker

Obtain 20 mL of a "normal urine" specimen and 20 mL of an "abnormal urine" sample. Your instructor will prepare all of the urine specimens. *Do not collect your own urine.* Describe the color of each specimen. Use pH paper or a pH meter to determine the pH of the urine samples. Determine the specific gravity of the urine sample with a urinometer. A display of urinometers may be set up in the lab. If a urinometer is not available, weigh a small beaker. Add 5 mL of the urine sample, and reweigh. Calculate the density and specific gravity of each sample.

B. Electrolytes

Materials: Test tubes, test tube rack, "urine" samples, flame test wire, Bunsen burner, small beakers, 3 M HCl, 0.1 M $AgNO_3$, 0.1 M $BaCl_2$, 3 M HNO_3, ammonium molybdate, warm water bath (70°C)

Sodium ion, Na^+ Dip a cleaned flame test wire (clean in 3 M HCl) into each "urine specimen" and place the wire loop in a flame. A bright, yellow-orange flame indicates the presence of sodium ion.

Chloride ion, Cl^- Place 3 mL of each "urine specimen" in a separate test tube. Add 5 drops of 3 M HNO_3, and 5 drops of 0.1 M $AgNO_3$. A white precipitate (AgCl) confirms the presence of chloride.

Sulfate ion, SO_4^{2-} Place 3 mL of each "urine specimen" in a separate test tube. Add 5 drops of 3 M HCl, and 5 drops of 0.1 M $BaCl_2$. A white precipitate ($BaSO_4$) confirms the presence of sulfate.

Phosphate ion, PO_4^{3-} Place 3 mL of each urine sample in a separate test tube. Add 5 drops of 3 M HNO_3 and 10 drops of ammonium molybdate solution. Place the test tubes in a warm water bath (70°C). A cloudy yellow precipitate confirms the presence of phosphate.

C. Glucose

Materials: Test tubes (2), test tube rack, "urine" samples, Benedict's solution, boiling water bath
Reagent strips: Clinistix, or multitest strips such as Multistix, Labstix, or Uristix

Reagent strips: Obtain Clinistix or a multitest strip and dip into a "urine specimen." Compare the test area with the color chart on the container or box. Report your observations. Give a quantitative evaluation of the glucose concentration.

Lab test (optional): A laboratory test for glucose uses Benedict's reagent. Place 10 drops of each "urine" sample in a separate test tube. Add 5 mL Benedict's reagent to each. Place the test tubes in a boiling water bath for 5 minutes. Cool. Record any changes in color. If glucose is present, estimate the amount.

Color with Benedict's Reagent	g/dL
Blue	<0.10
Blue-green	0.25
Green	0.50
Yellow	1
Orange	>2

D. Ketone Bodies

Materials: "Urine" samples; Ketostix, Labstix, or Multistix strips

Reagent strips: Obtain Ketostix, Labstix, or Multistix strips to test each urine specimen for ketone bodies. Dip the reagent strip into a "urine specimen." At 15 seconds, compare the test area with the color chart on the container or box. Report your results as negative or positive. If positive, indicate a small, moderate, or large amount of ketone bodies.

E. Protein

Materials: "Urine" samples; Albustix, Labstix, or Multistix strips; test tubes (2); test tube rack; 1 M HAc; test tube holder

Reagent strips: Obtain Albustix, Labstix, or Multistix for testing protein in the "urine" samples. Dip the strip into a urine specimen. Compare the test areas with the color chart on the container or box. Report your observations.

Lab test (optional): Place 5 mL of each urine sample in separate test tubes. Heat the solution to boiling for 1–2 minutes. If a precipitate forms, add 5 drops of 1 M HAc. Heat for 1 more minute. The formation of a white cloudy precipitate indicates the presence of protein. Report your observations.

F. Urobilinogen

Materials needed: "Urine" samples; Urobilistix, Labstix, or Multistix strips

Reagent Strips: Obtain two Urobilistix, Labstix, or Multistix for testing urobilinogen. Following directions on the package, dip the strip into the "urine specimens." Compare the test areas with the color chart on the container or box. Report your observations.

Report Sheet - Lab 42

Date _____   Name _____

Section _____   Team _____

Instructor_____

Pre-Lab Study Questions

1. What substances are normally found in urine?

2. Why are glucose, ketone bodies, or protein not normally detected in urine?

3. What test would indicate a problem with carbohydrate metabolism? Kidney failure?

A, B. Color, pH, Specific Gravity, and Electrolytes

Test	"Normal Urine Specimen"	"Abnormal Urine Specimen"
A. Color		
pH		
Specific gravity		

B. Electrolytes (Indicate absent –, present +, strongly present ++)		
Na^+		
Cl^-		
SO_4^{2-}		
PO_4^{3-}		

Report Sheet - Lab 42

Questions and Problems

Q.1 A patient's urine has a pH of 8. Would you consider this a normal pH of urine?

C–F. Glucose, Ketone Bodies, Protein, and Urobilinogen

For the following substances, indicate normal or abnormal (slight or severe).

	"Normal Urine Specimen"		"Abnormal Urine Specimen"	
	Color with reagent	Concentration	Color with reagent	Concentration
C. Glucose				
Benedict's test (optional)				
	Positive or negative?	Amount	Positive or negative?	Amount
D. Ketone bodies				
E. Protein				
Lab test (optional)				
F. Urobilinogen				

Questions and Problems

Q.2   A patient is a diabetic. What would you expect when a Clinitest strip is placed in the patient's urine?

Q.3   A Ketostix turns deep purple with a patient's urine. What could this indicate?

Q.4   If the patient in question 3 is on a low-carbohydrate diet, what advice would you give?

Appendix
Materials and Solutions

Standard Laboratory Materials

The equipment and chemicals listed throughout the appendix are the materials needed in the laboratory to perform the experiments in this lab manual. The amounts given are recommended for 20–24 students working in teams. The following equipment is expected to be in the laboratory lockers or available from the laboratory stock and will not be listed in each experiment.

Aspirators
Beakers (50–400 mL)
Balances (top loading or centigram)
Büchner filtration apparatus and filter paper
Bunsen burners
Burets (50 mL)
Buret clamps
Clay triangles
Containers for waste disposal
Crucible and cover
Distilled water (special faucet or containers)
Droppers
Evaporating dish
Flask, Erlenmeyer (125–250 mL)
Filter paper for funnels
Funnel
Glass stirring rods
Gloves
Graduated cylinders (5–250 mL)

Hot plates
Ice
Iron rings
Litmus paper
Meter sticks
pH paper
Ring stand
Rulers
Shell vials
Stirring rod, glass
Spatulas
Stoppers
Test tubes (6", 8")
Test tube rack
Thermometer
Tongs (crucible and beaker)
Watch glass
Wire gauze

Additional Materials Needed for Individual Experiments

The equipment and chemicals listed in this section must be supplied by the instructor or stockroom.. The amounts given are recommended for 20–24 students working in teams of two students

21 Properties of Organic Compounds
   A. Color, Odor, and Physical State *(May be a display in lab)*

1–2	Chemistry handbook	20 g	NaCl(*s*)
20 g	KI(*s*)	20 g	Benzoic acid(*s*)
20 mL	Toluene	20 mL	cyclohexane

   B. Solubility *(This may be an instructor demonstration.)*

20 g	NaCl(s)	20 mL	Toluene
20 mL	Cyclohexane		

   C. Combustion *(This may be an instructor demonstration.)*

30	Wood splints	10 g	NaCl(*s*)
10 mL	Cyclohexane		

D. Functional Groups
    10   Organic model kit or prepared models of organic compounds to observe

22 Structures of Alkanes

A. Structures of Alkanes, B. Isomers, C. Cycloalkanes, and D. Haloalkanes
    10   Organic model kits or prepared models
    2   Chemistry handbooks

23 Reactions of Hydrocarbons

A. Types of Hydrocarbons
    10   Organic model kits or prepared models

B. Combustion *(This may be an instructor demonstration.)*

5	Wooden splints	50 mL	Cyclohexane
50 mL	Cyclohexene	50 mL	Toluene

C. Bromine Test *(This may be an Instructor Demonstration)*

50 mL	Cyclohexane	50 mL	Cyclohexene
50 mL	Toluene	50 mL	Unknowns (use same compounds)
100 mL	1% Br_2 in CH_2Cl_2		

D. Potassium Permanganate ($KMnO_4$) Test

50 mL	Cyclohexane	50 mL	Cyclohexene
50 mL	Toluene	50 mL	Unknowns (use same compounds)
100 mL	1% $KMnO_4$		

E. Identification of Unknown
Unknowns of same substances

24 Alcohols and Phenols

A. Structures of Alcohols and Phenol
    10   Organic model kits or prepared models

Use for B. C. and D.

50 mL	Ethanol	50 mL	*t*-butyl alcohol (2-methyl-2-propanol)
50 mL	Cyclohexanol	50 mL	2-propanol
50 mL	20% phenol	2–4	Unknowns (Use same compounds)

C. Oxidation of Alcohols
100 mL 2% chromate solution

D. Ferric Chloride Test
100 mL 1% $FeCl_3$

E. Identification of Unknown
Unknowns of same substances

25 Aldehydes and Ketones

A. Structures of Some Aldehydes and Ketones
    10   Organic model kits or prepared models

B. Properties of Aldehydes and Ketones

50 mL	Acetone	50 mL	Benzaldehyde
50 g	Camphor	50 mL	Cinnamaldehyde,
50 mL	Vanillin	50 mL	Propionaldehyde (propanal)
50 mL	Cyclohexanone	50 mL	2,3-Butanedione
2	Chemistry handbooks	2–3	Unknowns (use above compounds)

C. Iodoform Test for Methyl Ketones (Test tubes from part B.3)
Use compounds from B

100 mL	10% NaOH	200 mL Iodine test reagent

D. Oxidation of Aldehydes and Ketones
Use compounds from B   500 mL Benedict's reagent
E. Identification of an Unknown
Unknowns of same substances

26 Types of Carbohydrates

A. Types of Carbohydrates and B. Disaccharides
10   Organic model kits or prepared models

27 Tests for Carbohydrates

A. Benedict's Test for Reducing Sugars
200 mL 2% starch   500 mL Benedict's reagent
Place 50 mL each in dropper bottles:
2% glucose   2% fructose   2% sucrose   2% lactose
50 mL Unknown solutions from above *Examples*: glucose, fructose, sucrose, and lactose

B. Seliwanoff's Test for Ketoses
200 mL 2% starch   100 mL Seliwanoff's reagent
50 mL   Unknown solutions from above *Examples*: glucose, fructose, sucrose, and lactose
Place 50 mL each in dropper bottles:
2% glucose   2% fructose   2% sucrose   2% lactose
50 mL Unknown solutions from above *Examples*: glucose, fructose, sucrose, and lactose

C. Fermentation Test
6   Fermentation tubes (or 6 small test tubes and 6 large test tubes),
200 mL 2% starch   40 g   Baker's yeast (fresh)
50 mL Unknown solutions from above
Examples: glucose, fructose, sucrose, and lactose
Place 50 mL each in dropper bottles:
2% glucose   2% fructose   2% sucrose   2% lactose
50 mL Unknown solutions from above *Examples*: glucose, fructose, sucrose, and lactose

D. Iodine Test for Polysaccharides
10   Spot plates
200 mL 2% starch   100 mL iodine reagent
Place 50 mL each in dropper bottles:
2% glucose   2% fructose   2% sucrose   2% lactose
50 mL Unknown solutions from above *Examples*: glucose, fructose, sucrose, and lactose

E. Hydrolysis of Disaccharides and Polysaccharides
10   Spot plate (or watch glasses)
100 mL 10% NaOH   100 mL 10% HCl
100 mL Iodine reagent   500 mL Benedict's reagent

Place 50 mL each in dropper bottles:
2% glucose   2% fructose   2% sucrose   2% lactose

F. Testing Foods for Carbohydrates
3–4   Sugar samples (refined, brown, "natural," powdered), honey
            Syrups (corn, maple, fruit), Foods with starches: cereals, pasta, bread
100 mL Seliwanoff's reagent   100 mL Iodine reagent
500 mL Benedict's reagent

28 Carboxylic Acids and Esters

A. Carboxylic Acids and Their Salts
100 mL 10% NaOH   100 mL 10% HCl
50 mL   Glacial acetic acid   50 g   Benzoic acid

B. Esters

10	Organic model sets		
200 mL	Glacial acetic acid	50 g	Salicylic acid
20 mL	Methanol	20 mL	1-Pentanol
20 mL	1-Octanol	50 mL	85% H_3PO_4 (dropper bottle)
20 mL	1-Propanol	20 mL	Benzyl alcohol

C. Hydrolysis of Esters

Place in dropper bottles:

100 mL	Methyl salicylate	100 mL 10% NaOH
100 mL	10% HCl	

29 Aspirin and Other Analgesics

A. Preparation of Aspirin

10	Pans or large beakers	50 g	Salicylic acid(*s*)
100 mL	Acetic anhydride	50 mL	85% H_3PO_4 in a dropper bottle

B. Testing Aspirin Products

3–4   Commercial brands of aspirin and buffered aspirin, purified aspirin (and/or crude aspirin)

20 g	Acetylsalicylic acid	100 mL	0.15% Salicylic acid
100 mL	1% $FeCl_3$	100 mL	10% NaOH
100 mL	10% HCl		

C. Analysis of Analgesics

Saran wrap, rubber band to fit beaker   60   micropipettes
10 spot plates

Dropper bottles containing 1% solutions in ethanol of the following:

1% aspirin	1% ibuprofen
1% acetaminophen	1% naproxen
1% caffeine	1% over the counter drugs,

400 mL solvent (Prepare by combining 300 mL ethyl acetate with 100 mL hexane)

10	TLC plate with silica gel	UV lamp (short wavelength 254 nm)

30 Lipids

A. Triacylglycerols

5   Organic model kits or prepared models

B. Physical Properties of Some Lipids and Fatty Acids

25 g	Lecithin	25 g	Stearic acid
25 g	Cholesterol	100 mL	Methylene chloride, CH_2Cl_2

Place 25 mL each in dropper bottles:

Oleic acid	Vitamin A
Olive oil	Safflower oil

C. Bromine Test for Unsaturation

5	Organic model kits or models	Samples from B
100 mL	1% Br_2 in CH_2Cl_2	

D. Preparation of Hand Lotion

Team project: Steps D.1, D.2, and D.3 may be prepared by different teams in the lab.

10	10-mL graduated cylinders	20	50-mL or 100-mL beakers
50g	stearic acid	15 g	cetyl alcohol
25 g	lanolin (anhydrous)	15 mL	(dropper) triethanolamine
25 mL	glycerin	100 mL	ethanol
100 mL	distilled water		fragrance (optional)
	commercial hand lotion products		

31 Glycerophospholipids and Steroids
A. Isolating Cholesterol from Egg Yolk

10	eggs	10	50-mL flasks
10	steam baths	10	100-mL beakers
10	short-stem funnels, glass wool	2	melting point apparatus
500 mL acetone			

B. Isolating Lecithin from Egg Yolk
Egg yolk residue from Part A

10	steam baths	10	100-mL beaker
2	melting point apparatus	500 mL ethyl ether	

32 Saponification and Soaps
A. Saponification: Preparation of Soap
Optional: Hot plate and a stirring bar   200 mL Ethanol
100 mL 20% NaOH   200 mL Saturated NaCl solution
10 pairs of disposable gloves
100 g   Solid fats: lard, coconut oil , solid shortening, coconut oil
100 mL Liquid vegetable oils, olive or other vegetable oil

B. Properties of Soaps and Detergents
50 g   Commercial soaps, lab-prepared soap (from part A), detergent
50 mL   Safflower oil   100 mL 1% $CaCl_2$
100 mL 1% $MgCl_2$ 100 mL 1% $FeCl_3$

33 Amines and Amides
A. Structure and Classification of Amines
10   Organic model kits or prepared models
B. Solubility of Amines in Water
Place in dropper bottles:
30 mL   Aniline   30 mL   Triethylamine
30 mL   *N*-Methylaniline   100 mL 10% HCl
C. Neutralization of Amines with Acids
Test tubes from part B   1   100 mL 10% HCl
D. Amides
5   Organic model kits   50 g   Acetamide
50 g   Benzamide
Place 100 mL each in dropper bottles:
10% NaOH   10% HCl

34 Synthesis of Acetaminophen
A. Synthesis of Acetaminophen
25 g   *p*-Aminophenol   10 mL   Acetic anhydride
100 mL   85% H_3PO_4 in dropper bottle
B. Isolating Acetanilide from an Impure Sample
50 g   Impure acetanilide

35 Plastics and Polymerization
A. Classification of Plastics
Samples of plastic items: Nylon, Teflon tape, Saran, Styrofoam cups, plastic cups, milk cartons, yogurt containers, buttle wrap, detergent bottles, soda bottles, etc.
40   small pieces of each type of plastic.
100 mL vegetable oil   100 mL ethanol
100 mL glycerin,   100 mLacetone,

401

B. Gluep and Slime®

10	10-mL, 50-mL graduated cylinders
50	Styrofoam cups 20 plastic sticks or spatulas
	plastic gloves

500 ml saturated borate solution 800 mL Elmer's glue
200 mL 4 % polyvinyl alcohol solution

C. Polystyrene

10	funnels	10	filter papers
10	hot plates	20	10- or 20-mL beaker
10	50-mL beakers	10	wood stick heavy duty aluminum foil
50 g	alumina	50 mL	Styrene
10 g	benzoyl peroxide (or an acne preparation which contains 5% or10% benzoyl peroxide),		

D. Nylon

10	50-mL, 100-mL beakers	10	forceps
10	10-mL and 50-mL graduated cylinders	10	metal spatulas
	gloves (must not dissolve in hexane)		

300 mL 50% aqueous ethanol solution   100 mL 6 M HCl, 6 M NaOH
100 mL acetone
250 mL Solution 1: 4% hexamethylenediamine and NaOH
        (dissolve 3.0 g $H_2N(CH_2)_6NH_2$ and 1.0 g NaOH in 50 ml of distilled water)
        If solid, place the reagent bottle in hot water to melt (mp 39°C).
250 mL Solution 2: 4% sebacoyl chloride, $ClCO(CH_2)_8COCl$, in hexane
        (dissolve 2.0 ml sebacoyl chloride in 50 ml hexane.)

36   Amino Acids

A. Amino Acids

    10   Organic model kits or prepared models

B.   Chromatography of Amino Acids

1 box	Plastic wrap	1 box	Whatman #1 filter paper (12 cm $\times$ 24 cm)
50	Toothpicks or capillary tubing	1	Drying oven (80°C)
2	Hair dryers (optional)	1	Stapler

Place 50 mL each in dropper bottles:

1% Alanine	1% Glutamic acid
1% Serine	1% Aspartic acid
1% Lysine	1% Phenylalanine

50 mL   1% Unknown amino acids (Use samples from above list)

Chromatography solvent
100 mL   0.5 M NH_4OH 200 mL isopropyl alcohol
0.2% Ninhydrin spray reagent (in ethanol or acetone)

37   Peptide and Proteins

A. Peptide Bonds and B. Structure of Proteins

    5   Organic model set

C. Denaturation of Proteins

Place in dropper bottle:
100 mL 1% egg albumin   Dissolve 1 g egg albumin in water to make 100 mL or
                                     students can make a fresh egg albumin solution by
                                       mixing the egg white from one egg with 200 mL of water
                                       and filtering the mixture through cheesecloth into a beaker.

100 mL   10% HNO_3 100 mL   10% NaOH
100 mL   95% ethanol   100 mL   1% $AgNO_3$

D. Isolation of Casein (Milk Protein)
    200 mL Nonfat milk   200 mL   10% acetic acid

E. Color Tests for Proteins
    10 g   Casein from part D
    Place 100 ml each in dropper bottles:
        1% Glycine   1% Tyrosine
        1% Gelatin   1% Egg albumin (See part C.)
    10% NaOH   50 mL   Conc. HNO_3
    1 can   0.2% Ninhydrin reagent   200 mL 5% $CuSO_4$

38 Enzymes

A. Effect of Enzyme Concentration
    10   Spot plate (or plastic sheets)
    10   Timer   1% starch (buffered to pH 7)
    200 mL Amylase preparation 100 mL 1% Glucose
    100 mL Iodine reagent   500 mL Benedict's reagent

B. Effect of Temperature
    Amylase preparation   200 mL 1% Starch
    100 mL Iodine test reagent

C. Effect of pH
    Amylase preparation   100 mL buffers (pH 2, 4, 7, 10)
    200 mL 1% Starch   100 mL Iodine test reagent

D. Inhibition of Enzyme Activity
    Amylase preparation
    Place 50 mL each in dropper bottles:
        1% NaCl   1% $AgNO_3$
        1% $CuSO_4$ 1% $Pb(NO_3)_2$
        1% $HgCl_2$ Ethanol
    200 mL 1% Starch   100 mL Iodine reagent

39 Vitamins

A. Solubility of Vitamins
    4–6   Samples of vitamins A, B, C, D, E, folic acid, or others
    50 mL Methylene chloride (CH_2Cl_2)

B. Standardization of Vitamin C
    2   Mortar and pestle   10   Tablets of vitamin C (100-mg)
    100 mL 0.1 *M* HAc   100 mL 1% Starch
    1 L   iodine solution

C. Analysis of Vitamin C In Fruit Juices and Fruit Drinks
    100 mL 0.1 *M* $HC_2H_3O_2$ 100 mL 1% Starch
    1 L   Iodine solution   10 squares of cheesecloth
        Samples of fruit juices/drinks: orange, grapefruit, drinks such as HiC or Snapple, or powdered drinks such as Tang or Kool-Aid, or vegetables pureed in a blender or juicer

D. Heat Destruction of Vitamin C
    10   Squares of cheesecloth   100 mL 1% Starch
    100 mL 0.1 *M* HAc   1 L   Iodine solution
    3–4   Fruit juices/drinks: orange, grapefruit, drinks such as HiC or Snapple, or powdered drinks such as Tang or Kool-Aid, or vegetables pureed in a blender or juicer

40 DNA Components and Extraction

A. Components of DNA

10 organic model kits

B. Extraction of DNA

Different student teams may extract DNA from different DNA sources and compare results or the procedure may be run initially with onions and then repeated using other DNA sources selected by each student team.

DNA sources: white onions, cauliflower, broccoli, garlic, split peas, bananas, and/or animal sources such as chicken liver, calf thymus

5	knives	2	blenders
	cheesecloth or filter paper, ice	1-2	microscopes

1 L SDS-NaCl-EDTA buffer solution (50 g sodium dodecyl sulfate, 50 g NaCl, 5 g sodium citrate, 0.2 mL of 0.5 EDTA and water to make 1 L)

500 mL isopropanol (in an ice bath)

41 Digestion of Foods

A. Digestion of Carbohydrates

10	Spot plates (or plastic sheets)	Amylase preparation
200 mL	1% Starch	100 mL Iodine reagent
500 mL	Benedict's reagent	

B. Digestion of Fats

2–3	pH meters or pH paper	1 qt	Whole milk
200 mL	0.1 M NaOH	50 mL	Safflower oil
50 mL	Bile salts	100 mL	2% pancreatin
100 mL	1% Phenolphthalein		

C. Protein Digestion

2–3	Hard-boiled eggs	100 mL	2% pepsin
100 mL	0.1 M HCl		

42 Analysis of Urine

A. Color, pH, and Specific Gravity

3 Urinometers (hydrometers), *or* a small graduated cylinder and beaker

100 mL "Normal Urine" specimen   100 mL "Abnormal Urine" specimen

B. Electrolytes

10 Flame test wires   100 mL 3 M HCl

Place 100 mL in dropper bottles:

0.1 M AgNO$_3$ 0.1 M BaCl$_2$

3 M HNO$_3$ (NH$_4$)$_2$MoO$_4$ ammonium molybdate reagent

C. Glucose

20 Reagent strips: Clinistix, or multitest strips such as Multistix, Labstix, or Uristix

100 mL "Normal Urine"

100 mL "Abnormal Urine"

100 mL Benedict's reagent

D. Ketone Bodies

20 Reagent strips: Clinistix, or multitest strips such as Multistix, Labstix, or Uristix

100 mL "Normal Urine"

100 mL "Abnormal Urine"

E. Protein
    20   Ketostix, Albustix, Labstix, or Multistix strips
    100 mL 1 M HAc
    100 mL "Normal Urine"
    100 mL "Abnormal Urine"

F. Urobilinogen
    20   Urobilistix, Labstix, or Multistix strips
    100 mL "Normal Urine"
    100 mL "Abnormal Urine"

Preparation of Solutions Used in the Laboratory

Acids and bases

Acetic acid $C_2H_3O_2$ (HAc)

0.1 M HAc	Dilute 0.6 mL of glacial acetic acid with water to make 100 mL
1 M HAc	Dilute 6 mL of glacial acetic acid with water to make 100 mL
10% HAc	Dilute 50 mL of glacial HAc with water to make 500 mL

Ammonium hydroxide NH_4OH

0.1 M NH_4OH	Dilute 6.7 mL conc. NH_4OH with water to make 1.0 L
0.5 M NH_4OH	Dilute 34 mL conc. NH_4OH with water to make 1.0 L
1 M NH_4OH	Dilute 67 mL conc. NH_4OH with water to make 1.0 L

Hydrochloric acid HCl

0.1 M HCl	Dilute 8.3 mL of conc. HCl with water to 1 L and standardize against standardized 0.1 M NaOH (or 0.2 M NaOH)
1.0 HCl	Dilute 85 mL conc. HCl with water to make 1.0 L
2.0 M HCl	Dilute 170 mL conc. HCl with water to make 1.0 L
3.0 M HCl	Dilute 250 mL conc. HCl with water to make 1.0 L
10% HCl	Dilute 230 mL con. HCl to make 1.0 L

Nitric acid HNO_3

6 M HNO_3	Dilute 76 mL conc. HNO_3 with water to make 200 mL
10% HNO_3	Dilute 10 mL conc. HNO_3 with water to make 100 mL

Sodium hydroxide NaOH

0.1 M NaOH   Dissolve 4.0 g NaOH in water to make 1.0 L
Standardization: Weigh a 1-g sample of potassium hydrogen phthalate, $KC_8H_5O_4$, to 0.001 g. Dissolve in 25 mL of water, add phenolphthalein indicator, and titrate with the prepared NaOH solution. Calculate the molarity (3 significant figures) as

$$\text{g phthalate} \times \frac{1 \text{ mole phthalate}}{204 \text{ g phthalate}} \times \frac{1}{\text{L NaOH used}} = \underline{\quad} M$$

6 M NaOH	Dissolve 240 g NaOH in water to make 1.0 L
3 M NaOH	Dissolve 120 g NaOH in water to make 1.0 L
10% NaOH	Dissolve 10 g NaOH in water to make 100 mL
20% NaOH	Dissolve 20 NaOH in water to make 100 mL

Salt solutions

Aluminum sulfate	1% $Al_2(SO_4)_3$	Dissolve 1.5 g $Al_2(SO_4)_3 \cdot 9H_2O$ in water to make 100 mL
Ammonium chloride	0.1 M NH_4Cl	Dissolve 0.54 g NH_4Cl in water to make 100 mL

Ammonium molybdate — Dissolve 8.1 g H_2MoO_4 in 20 mL water. Add 6 mL conc. NH_4OH to give a saturated solution. Filter. Slowly add filtrate to a mixture of 27 mL conc. HNO_3 and 40 mL water. Let stand 1 day. Filter and add water to 100 mL.

Ammonium oxalate	0.1 M $(NH_4)_2C_2O_4$	Dissolve 1.4 g $(NH_4)_2C_2O_4 \cdot H_2O$ with water to 100 mL
Barium chloride	0.1 M $BaCl_2$	Dissolve 2.4 g $BaCl_2 \cdot 2H_2O$ in water to make 100 mL
Barium nitrate	0.1 M $Ba(NO_3)_2$	Dissolve 2.6 g $Ba(NO_3)_2$ in water to make 100 mL
Calcium chloride	1% $CaCl_2$	Dissolve 1.3 g $CaCl_2 \cdot 2H_2O$ in water to make 100 mL
	20% $CaCl_2$	Dissolve 200 g $CaCl_2$ in water to make 1.0 L
	0.1 M $CaCl_2$	Dissolve 1.5 g $CaCl_2 \cdot 2H_2O$ in water to make 100 mL
Copper(II) chloride	0.1 M $CuCl_2$	Dissolve 1.7 g $CuCl_2 \cdot 2H_2O$ in water to make 100 mL
Copper(II) sulfate	0.1 M $CuSO_4$	Dissolve 2.5 g $CuSO_4 \cdot 5H_2O$ in water to make 100 mL
	1 M $CuSO_4$	Dissolve 25 g $CuSO_4 \cdot 5H_2O$ in water to make 100 mL
	1% $CuSO_4$	Dissolve 0.78 g $CuSO_4 \cdot 5H_2O$ in water to make 50 mL
	5% $CuSO_4$	Dissolve 15.6 g $CuSO_4 \cdot 5H_2O$ in water to give 200 mL
Iron(III) chloride	0.1 M $FeCl_3$	Dissolve 2.7 g $FeCl_3 \cdot 6H_2O$ in water to make 100 mL
	1% $FeCl_3$	Dissolve 1.7 g $FeCl_3 \cdot 6H_2O$ in water to give 100 mL
Iron(III) nitrate	0.01 M $Fe(NO_3)_3$	Dissolve 0.40 g $Fe(NO_3)_3 \cdot 9H_2O$ in water to make 100 mL
	1 M $Fe(NO_3)_3$	Dissolve 40. g $Fe(NO_3)_3 \cdot 9H_2O$ in water to make 100 mL
Lead(II) nitrate)	1% $Pb(NO_3)_2$	Dissolve 0.5 g $Pb(NO_3)_2$ in water to make 50 mL
Magnesium chloride	1% $MgCl_2$	Dissolve 2.1 g $MgCl_2 \cdot 6H_2O$ in water to make 100 mL
Mercury(II) chloride	1% $HgCl_2$	Dissolve 0.5 g $HgCl_2$ in water to make 100 mL
Potassium chloride	0.1 M KCl	Dissolve 0.75 g KCl in water to make 100 mL
Potassium permanganate	1% $KMnO_4$	Dissolve 1g $KMnO_4$ in water to give 100 mL
Potassium thiocyanate	0.01 M $KSCN$	Dissolve 0.097 g $KSCN$ in water to make 100 mL
	0.1 M $KSCN$	Dissolve 0.97 g $KSCN$ in water to make 100 mL
	1 M $KSCN$	Dissolve 9.7 g $KSCN$ in water to make 100 mL
Silver nitrate	0.1 M $AgNO_3$	Dissolve 1.7 g $AgNO_3$ in water to make 100 mL
	1% $AgNO_3$	Dissolve 0.5 g $AgNO_3$ in water to make 50 mL
Sodium carbonate	0.1 M Na_2CO_3	Dissolve 2.9 g $Na_2CO_3 \cdot 7H_2O$ in water to make 100 mL
Sodium chloride	0.1 M $NaCl$	Dissolve 0.58 g $NaCl$ in water to make 100 mL
	1% $NaCl$	Dissolve 1 g $NaCl$ in water to make 100 mL
	10% $NaCl$	Dissolve 10 g $NaCl$ in water to make 100 mL
	20% $NaCl$	Dissolve 200 g $NaCl$ in water to make 1.0 L
	Saturated $NaCl$	Add 80 g $NaCl$ to water to make 200 mL
Sodium phosphate	0.1 M Na_3PO_4	Dissolve 3.8 g $Na_3PO_4 \cdot 12H_2O$ in water to make 100 mL

Sodium sulfate	0.1 M Na_2SO_4	Dissolve 3.2 g $Na_2SO_4 \cdot 10H_2O$ in water to make 100 mL
	1% Na_2SO_4	Dissolve 2.3 g $Na_2SO_4 \cdot 10H_2O$ in water to make 100 mL
Strontium chloride	0.1 M $SrCl_2$	Dissolve 1.95 g $SrCl_2 \cdot 2H_2O$ in water to make 100 mL

Carbohydrates

Fructose	2% fructose	Add 1 g fructose to water to make 50 mL
Glucose	0.1 M glucose	Dissolve 1.8 g glucose in water to make 100 mL
	2% glucose	Add 1 g glucose to water to make 50 mL
	10% glucose	Dissolve 10 g glucose in water to make 100 mL
Lactose	2% lactose	Add 1 g lactose to water to make 50 mL
Sucrose	0.1 M sucrose	Dissolve 3.42 g sucrose in water to make 100 mL
	2% sucrose	Add 1 g sucrose to water to make 50 mL
Starch	1%	Make a paste of 2 g soluble starch and 40 mL water. Add to 160 mL of boiling water to make 200 mL. Stir and cool.
	2%	Make a paste of 4 g soluble starch and 40 mL water. Add to 160 mL of boiling water to make 200 mL. Stir and cool.

Reagents

Benedict's reagent — Dissolve 86 g sodium citrate, $Na_3C_6H_5O_7$, and 50 g anhydrous Na_2CO_3 in 400 mL water. Warm. Dissolve 8.6 g $CuSO_4 \cdot 5H_2O$ in 50 mL water. Add to sodium citrate solution, stir, and add water to make 500 mL solution.

DNA buffer — SDS-NaCl-EDTA buffer solution (50 g sodium dodecyl sulfate, 50 g NaCl, 5 g sodium citrate, 0.2 mL of 0.5 EDTA and water to make 1 L)

0.15% Salicylic acid — Dissolve 0.15 g salicylic acid in 100 mL water.
Seliwanoff's reagent — Dissolve 0.15 g resorcinol in 100 mL 6 M HCl

Iodine solution — Dissolve 10 g I_2 + 20 g KI in water to make 500 mL
Iodoform (iodine solution) — Dissolve 10 g I_2 + 20 g KI in water to make 100 mL

0.2% Ninhydrin — Dissolve 0.2 g in ethanol to make 100 mL
20% Phenol — Dissolve 20 g phenol in water to make 100 mL

4% Hexamethylenediamine/NaOH
Combine 3.0 g $H_2N(CH_2)_6NH_2$ + 1.0 g NaOH and add water to 50 mL

4% Sebacoyl chloride/Hexane
Combine 20.0 mL $ClCO(CH_2)_8COCl$ and hexane to make 50 mL

Indicators

1% Phenolphthalein — Dissolve 1 g phenolphthalein in 50 mL ethanol and 50 mL water
1% Bromine solution — Dilute 1 mL Br_2 with methylene chloride to make 100 mL
2% Chromate — Dissolve 2.0 g $K_2Cr_2O_7$ in 10 mL of 6 M H_2SO_4; then carefully add to water to make 100 mL

Amino acids

1% Alanine — Dissolve 0.5 g alanine in water to make 50 mL
1% Aspartic acid — Dissolve 0.5 g aspartic acid in water to make 50 mL
1% Glutamic acid — Dissolve 0.5 g glutamic acid in water to make 50 mL
1% Glycine — Dissolve 1 g glycine in water to make 100 mL

1% Lysine	Dissolve 0.5 g lysine in water to make 50 mL
1% Phenylalanine	Dissolve 0.5 g phenylalanine in water to make 50 mL
1% Serine	Dissolve 0.5 g serine in water to make 50 mL
1% Tyrosine	Dissolve 1 g tyrosine in water to make 100 mL

Proteins

1% Egg albumin	Dissolve 1 g egg albumin in water to make 100 mL
1% Gelatin	Dissolve 1 g gelatin in water to make 100 mL
2% Pancreatin	Dissolve 2.0 pancreatin in 100 mL 0.25% Na_2CO_3. Use immediately.
2% Pepsin	Dissolve 2% pepsin in water to make 100 mL

Urine specimens

| "Normal Urine" | Mix 2 mL of 0.1 M NaCl, 2 mL of 0.1 M Na_2SO_4, 2 mL of 0.1 M Na_3PO_4, Add 1 g of urea. Adjust pH to 5.5–7.0 with 5 to 10 drops of 6 M HCl and dilute to 100 mL. |
| "Abnormal Urine" | To the recipe for "normal urine," add 1 g glucose, 1 g egg albumin, 4 mL acetone, and 1 mL 6 M HCl. |